Model-Based Predictive Control of Electric Drives

Arne Linder
Rahul Kanchan
Ralph Kennel
Peter Stolze

Cuvillier Verlag Göttingen

Bibliografische Information der Deutschen Nationalbibliothek
Die Deutsche Nationalbibliothek verzeichnet diese Publikation in der Deutschen Nationalbibliografie; detaillierte bibliografische Daten sind im Internet über http://dnb.d-nb.de abrufbar.

1. Aufl. - Göttingen: Cuvillier, 2010

978-3-86955-398-6

Nonnenstieg 8, 37075 Göttingen
Telefon: 0551-54724-0
Telefax: 0551-54724-21
www.cuvillier.de

1. Auflage, 2010
Gedruckt auf säurefreiem Papier

978-3-86955-398-6

Contents

List of Figures

List of Tables

Preface

This book is a translation of the PhD thesis submitted by Arne Linder to the Department of Electrical, Information and Printing Technology at Wuppertal University in 2005. Dr. Arne Linder showed in his outstanding work for the first time that in general model-based predictive control of electrical drives, even with prediction horizons greater than one sampling cycle is possible, despite the high sampling rates which are necessary in drive control.

The presented work shall give a general overview about the predictive control methods used so far in drive control as well as a classification of them. Furthermore, a family tree of predictive control methods in drive technology is presented which also shows relations between them, derivations and further development of certain methods. Generalized Predictive Control (GPC) is also introduced, first the two PI current controllers in a field-oriented control scheme of an induction machine are replaced by two single GPC controllers and their results are then compared to the ones of classical PI controllers. In a next step the speed controller is replaced by a GPC contoller. Different types of models, methods for handling nonlinearities and model discretization are also discussed in this work. The ability of GPC to provide multivariable control is examined, using one single controller for both flux- and torque-producing currents. Moreover, possibilities for a direct inverter control method which makes the modulator needless are presented as well: The chances of this approach, problems but also possible remedies are discussed in detail. As a conclusion further control methods which are partly considered to be predictive are discussed. The sections contain not only simulation, but also experimental results which clearly prove the applicability of the presented control methods for electrical drives.

I would like to thank Dr. Rahul Kanchan, Dipl.-Ing. Peter Stolze and Dipl.-Ing. (syr.) Nael ALSheakh Ameen for the translation of this work. M.Sc. Juan Carlos Ramirez Martínez, M.Sc. Bahaa Galal and M.A. Johanna Schmidt helped to finalize this work by proofreading the English translation.

Munich, June 2010 Ralph Kennel

1 Introduction

In industry, variable speed electrical drives are applied to various applications; their power bandwidth differs from a few watts for small servo-motors up to several hundreds of kilowatts for traction applications. For decades, the DC series motor was the mostly used drive motor because of its simple speed controllability; for smaller power demands, permanent-magnet DC machines were also used. Nevertheless, the disadvantage associated with these machines is the wear-associated commutator which leads to a rise in maintenance expenses and also to a raised soiling of the machine by brush abrasion. At the same time, a complete encapsulation of all electrical components is relatively complex due to the construction. Rotating field AC machines are known since the initial times of electrical machines, however, they were rarely used as the speed of these machines changes with the frequency of the supplying grid. Consequently, machines of this kind were only used where this feature was required, e. g. as generators in power stations, or, e. g., for fans where speed adjustment is unnecessary.

Along with the advances of semiconductor technology, the possibilities to make AC machines in a simple way speed variable increased. Therefore, the supply voltage is rectified into a so-called DC link, from where a three-phase voltage with variable frequency can be generated with the help of an inverter. Herewith, the task to operate the machine with any speed was solved, however, a dynamic control similar to DC machines was not possible. This was due to the fact that the separate control of field flux and torque was not available. This problem was solved with the introduction of the so-called field-oriented control (see chapter 2) about 30 years ago.

Since this time, field-oriented control for synchronous as well as for asynchronous machines is state-of-the-art; thereby, PI controllers are used in a cascaded structure (see chapter 3). Undesirable side effects and nonlinearities of the machine are tackled with precontrols or feedforward compensation techniques in such a way that the quality of the achieved control meets the requirements of all drive applications existing so far.

Now the question rises why new control techniques should be examined if the control techniques available today can fulfill all known requirements until now.

This shall be explained on the basis of the classical structure of a controlled electrical drive. As shown in figure 1.1 a position control scheme consists of a threefold cascaded control structure. The external control loops consist of an integrator which describes the behavior of the system's inertia and the gearbox, respectively, and the next inner control loop. When designing such a cascaded structure normally at first the inner loop is approximated by a PT_1-block and then the controller parameters are determined according to the so-called *symmetrical optimum*. However, good control properties can only be achieved if the time constants of the inner control loop and the subsequent external control loop differ by a factor of at least 7–10 [38]. Consequently, for a position control scheme, the current controller must be about 50–100 times faster than the position control loop. It can be foreseen that, sooner or later, the dynamics which can be achieved with a cascade structure are no longer suitable for highly dynamic drive applications as a sufficiently fast current controller is not feasible. As the need for an ultra fast current control loop arises solely from the cascaded structure itself, the desired fast dynamics in the position or speed control loop could be achieved with other control structures, even if the computation time of the control process is significantly higher than with cascaded control. Hence, it can be seen that, in the medium term, linear controllers in cascaded structure can no longer meet the growing requirements for the dynamics of electrical drives. Predictive or precalculating controllers which need no cascaded control offer an alternative. Chapter 4 explains the basic principle of such a control strategy.

All control strategies published so far in the domain of drive technology precalculate the control behavior only for the next sampling cycle. Correspondingly, the optimization of the actuating variable can only be done for this single precalculated sampling cycle. However, from classic control theory, strategies are known which allow a higher prediction horizon and are therefore called *Long-Range Predictive Control*. Since they use a model of the controlled process for the prediction, they are also called *model predictive controllers* (MPC) or *model-based predictive controllers*. These strategies are discussed in detail in chapter 5. Due to the longer prediction horizon, these strategies are relatively complex in their calculations in comparison to ordinary linear controllers; hence, their implementation in drive technology was not possible yet. Nevertheless, the availability of faster microprocessors and computing devices is increasing continuously and at the same time their price is continuously decreasing. So the argument of too high computational expenses has already lost much of its relevance and will lose even more of it in the future. Hence, the implementation of even complex model-based strategies can be expected to be

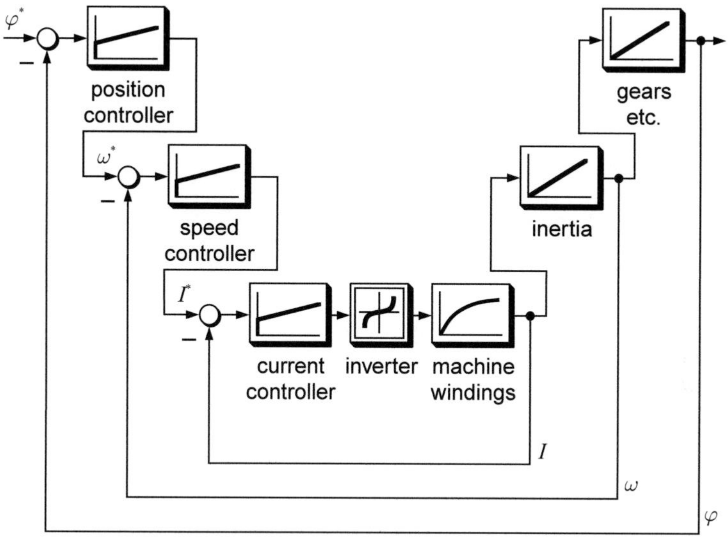

Figure 1.1: Typical structure of a cascade controller

realized in some years. In the chapters 6–9 different MPC strategies are introduced and their advantages and disadvantages are explained with respect to their implementation in drive technology.

2 Field-oriented control

In drive applications a DC machine has the great advantage that it has separate field and armature windings. Thus both of its most important states, namely flux and torque, can be controlled independently of each other by the field winding current and by the armature current, respectively. The commutator acting as a mechanical converter makes sure that the current distribution of the rotor is always positioned properly, so that the rotor current and the main flux are always in quadrature to each other. Asynchronous machines, on the other hand, which offer great advantages because of their maintenance free operation, have only one active winding, the stator winding. The rotor winding is either short circuited (squirrel-cage motor) or connected with variable resistors (slip ring rotor). Consequently, both flux and torque in the machine have to be raised via the stator windings. Splitting the control structure in a simple way into flux and torque control is not possible. However, for high drive performance, both states must be controlled independently of each other [57].

A remedy for solving this problem is the use of the so-called *field-oriented control* [79], which can be described with the help of *space vector representation* [70]. The basic idea behind this representation is that the three-phase current system can be represented by a three-axis coordinate system as shown in figure 2.1(a). Unfortunately, the three axes a, b and c are not linearly independent of each other, a fact that complicates a mathematical description of the actions in a three-phase system. Hence, an alternative two-phase system with two axes linearly independent of each other is constructed. Figure 2.1(b) shows this equivalent system. To obtain an easy representation of the two phase quantities, usually a complex coordinate system is selected. Thus, the corresponding quantities from one coordinate system, e. g. the stator voltage quantities, can be transformed into another coordinate system with the help of the transformation equations (2.1) to (2.4).

$$\boldsymbol{u}_s = \tfrac{2}{3}\left(u_{sa} + \boldsymbol{a}\cdot u_{sb} + \boldsymbol{a}^2\cdot u_{sc}\right) \qquad \text{in which} \qquad \boldsymbol{a} = e^{j\frac{2\pi}{3}} \tag{2.1}$$

$$u_0 = \tfrac{1}{3}\left(u_{sa} + u_{sb} + u_{sc}\right) \tag{2.2}$$

$$u_{s\alpha} = \mathfrak{Re}\{\boldsymbol{u}_s\} = \tfrac{2}{3}\left(u_{sa} - \tfrac{1}{2}\,u_{sb} - \tfrac{1}{2}\,u_{sc}\right) \tag{2.3}$$

$$u_{s\beta} = \mathfrak{Im}\{\boldsymbol{u}_s\} = \tfrac{1}{\sqrt{3}}\left(u_{sb} - u_{sc}\right) \tag{2.4}$$

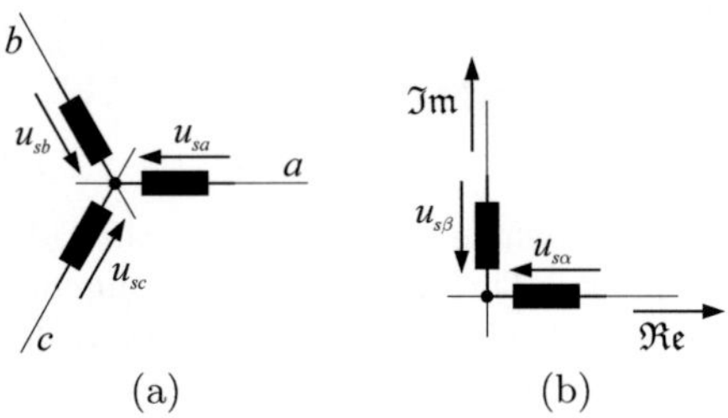

Figure 2.1: Three-phase system and alternative two-phase system

The idea of field-oriented control is based on the fact that the three stator currents i_{sa}, i_{sb} and i_{sc} in the three stator windings are first transformed into a complex state space vector $\boldsymbol{i}_s$. Therefore, a transformation analog to (2.1) is used for the stator currents. Then, an induction machine can be described in an arbitrary reference coordinate system, rotating with ω_k, with the following equations [70, 71]:

$$\boldsymbol{u}_s = r_s \cdot \boldsymbol{i}_s + \frac{d\boldsymbol{\psi}_s}{d\tau} + j\omega_k\boldsymbol{\psi}_s \tag{2.5}$$

$$0 = r_r \cdot \boldsymbol{i}_r + \frac{d\boldsymbol{\psi}_r}{d\tau} + j(\omega_k - \omega)\boldsymbol{\psi}_r \tag{2.6}$$

$$\boldsymbol{\psi}_s = l_s \cdot \boldsymbol{i}_s + l_h \cdot \boldsymbol{i}_r \tag{2.7}$$

$$\boldsymbol{\psi}_r = l_r \cdot \boldsymbol{i}_r + l_h \cdot \boldsymbol{i}_s \tag{2.8}$$

For current control, as it is implemented in the internal loop of a cascade control structure, it is an adequate solution to choose $\boldsymbol{i}_s$ und $\boldsymbol{\psi}_r$ as state variables. After reformulating according to [56, 57], the following differential equations can be obtained:

$$\boldsymbol{i}_s + \tau_\sigma{}'\frac{d\boldsymbol{i}_s}{d\tau} = -j\omega_k\tau_\sigma{}'\boldsymbol{i}_s + \frac{k_r}{r_\sigma}\left(\frac{1}{\tau_r} - j\omega\right)\boldsymbol{\psi}_r + \frac{1}{r_\sigma}\boldsymbol{u}_s \tag{2.9}$$

$$\boldsymbol{\psi}_r + \tau_r\frac{d\boldsymbol{\psi}_r}{d\tau} = -j(\omega_k - \omega)\tau_r\boldsymbol{\psi}_r + l_h \cdot \boldsymbol{i}_s \tag{2.10}$$

in which $\tau_s = \frac{l_s}{r_s}, \tau_r = \frac{l_r}{r_r}, \sigma = 1 - \frac{{l_h}^2}{l_s l_r}$, $l_s{}' = \sigma l_s$, $l_r{}' = \sigma l_r$, $k_r = \frac{l_h}{l_r}$, $k_s = \frac{l_h}{l_s}$, $\tau_s{}' = \frac{\sigma l_s}{r_s}$, $\tau_r{}' = \frac{\sigma l_r}{r_r}$, $r_\sigma = r_s + r_r \cdot {k_r}^2$ and $\tau_\sigma{}' = \frac{\sigma l_s}{r_\sigma}$. Figure 2.2 shows the corresponding signal flow graph. In this work, double lines represent complex values.

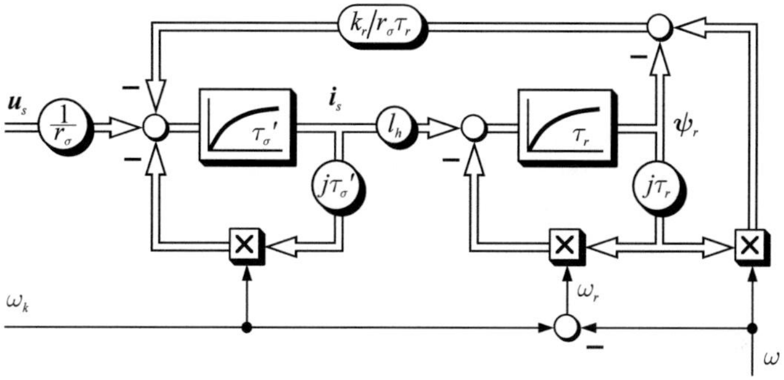

Figure 2.2: Complex machine model

Now a coordinate system rotating with the speed of the rotating field is selected as the base coordinate system with the real axis aligned in the direction of the field. With the help of this new coordinate system the stator current space vector $\boldsymbol{i}_s$ can be divided into a torque-producing component i_{sq} and into a flux-producing component i_{sd}. As both of these current components can be controlled independently of each other, dynamic control similar to a shunt-wound DC machine can be achieved. If the complex equations (2.9) and (2.10) are divided into their real and imaginary parts, four scalar equations which describe the dynamic behavior of the induction machine can be obtained. Because of field orientation $\omega_k = \omega_s$ is set; for the same reason $\psi_{rq} = 0$ is set. The appropriate signal flow graph can be seen in figure 2.3.

$$i_{sd} + \tau_\sigma{}' \frac{di_{sd}}{d\tau} = \omega_s \tau_\sigma{}' i_{sq} + \frac{k_r}{r_\sigma \tau_r} \psi_{rd} + \frac{1}{r_\sigma} u_{sd} \tag{2.11}$$

$$i_{sq} + \tau_\sigma{}' \frac{di_{sq}}{d\tau} = -\omega_s \tau_\sigma{}' i_{sd} - \frac{k_r}{r_\sigma} \omega \psi_{rd} + \frac{1}{r_\sigma} u_{sq} \tag{2.12}$$

$$\psi_{rd} + \tau_r \frac{d\psi_{rd}}{d\tau} = l_h i_{sd} \tag{2.13}$$

$$0 = -(\omega_s - \omega) \tau_r \psi_{rd} + l_h i_{sq} \tag{2.14}$$

Equation (2.14) is of no significance for the control and hence, it is not considered for further analysis. However, it describes the condition for field orientation, i. e. the position angle of the field coordinate system with respect to the fixed stator coordinate frame can be determined with it. Therefore equation (2.14) is resolved for the slip or rotor frequency[1] $\omega_r = \omega_s - \omega$ and after doing this, the following equation can be obtained:

$$\omega_r = (\omega_s - \omega) = \frac{l_h i_{sq}}{\tau_r \psi_{rd}} \tag{2.15}$$

The value of ψ_{rd} required for the calculation of the above relationship can be obtained from the differential equation (2.13). By adding the calculated slip speed ω_r to the mechanical rotating speed ω of the rotor, the stator speed ω_s results. The integration of ω_s provides the field angle δ. The overall signal flow graph of the induction machine can be seen in figure 2.4.

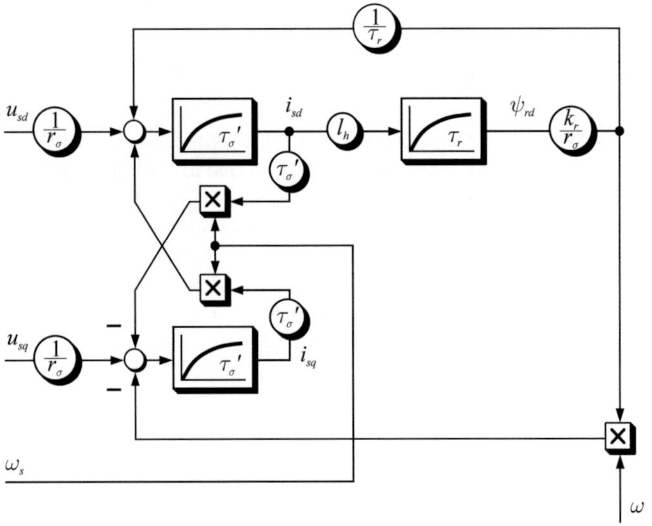

Figure 2.3: Scalar machine model (field coordinates)

[1] The rotor frequency is *not* the mechanical rotating frequency but the frequency of the currents flowing in the rotor.

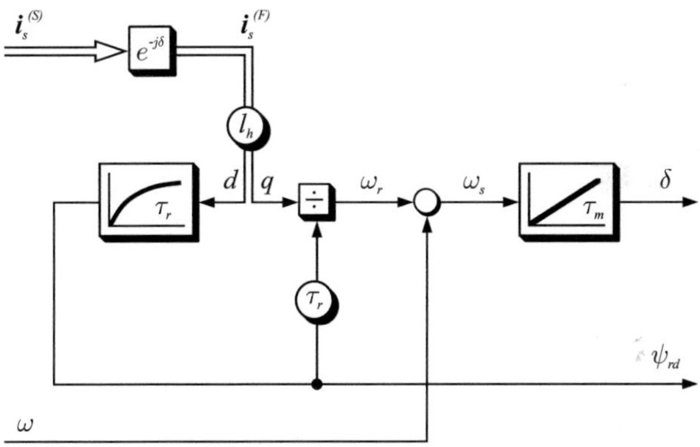

Figure 2.4: Rotor model

3 Cascade control with PI controllers

Before investigating in predictive control of induction machines, it is meaningful to create a basic control scheme for comparisons first. Therefore, field-oriented control of an induction machine with conventional PI controllers for current and speed is realized in a cascaded structure.

Conventional PI controllers can only control one actuating variable with the help of one reference variable. Controllers of this type are called SISO (single input, single output) controllers. In contrast to this, there are controller structures which can control several actuating variables at the same time; these are called MIMO (multiple input, multiple output) controllers or multidimensional controllers. Thus, if in an electrical drive the current as well as the speed has to be controlled with PI controllers, the control structure must be implemented as a cascaded control. Figure 3.1 shows the typical structure of field-oriented control with cascaded PI controllers. The internal control loop is formed by both current controllers for the field- and for the torque-producing stator current components. In order to improve the clarity of the signal flow graph both current controllers are represented by one single complex variable current controller. The speed and flux control loops are superordinated.

3.1 Current control

For the realization of the current control, at first the stator equations (2.11) and (2.12) are considered. Since current control has to be implemented with simple PI controllers, it makes sense to neglect the effect of the flux ψ_{rd} on the currents as well as to neglect the cross coupling between the two current components in equation (2.9). Figure 3.2 shows the resulting structure. The actual parameters of the induction machine used for the investigations in this book are given in appendix D.

Apart from the time constants of the stators windings, the dead time of the inverter must be taken into consideration while designing the PI controller; it is approximated with another PT_1-block with the time constant $\tau_{tot} = 1.5\,\tau_0$. The complete control loop for the flux-producing current component can be seen in figure 3.3. Since both current control loops for the field- and for the

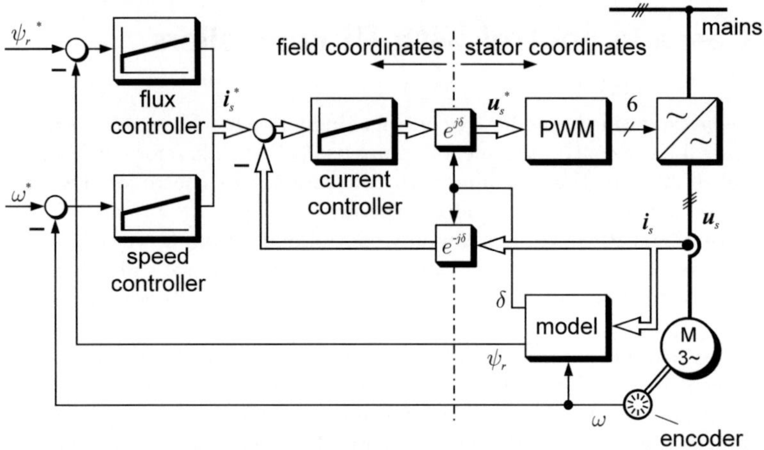

Figure 3.1: Field-oriented drive control with PI controllers

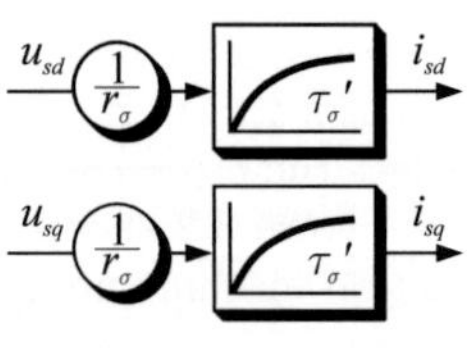

Figure 3.2: Simplified model for current control (field coordinates)

torque-producing current components are identical, the signal flow graph for the control of i_{sq} is the same as the one for i_{sd}. The two effective time constants τ_{tot} and $\tau_\sigma{}'$ differ from each other by a factor of about 50, thus the controller has to be designed according to the symmetric optimum [38].

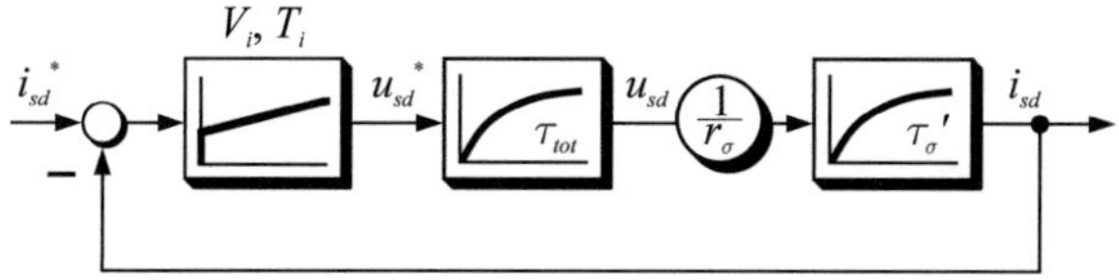

Figure 3.3: Current control loop with PI controller

3.2 Speed control

The design of the outer speed control loop requires the approximation of the behavior of the internal current control loop with a first order transfer function. Thereby, the time constant τ^* describes the dynamics of the complete inner current control loop. The mechanical time constant of the experiment setup can be determined experimentally. Since the speed feedback signal bounces in discrete steps if a low resolution incremental encoder with relative high quantization error is used, it must be smoothed with a low-pass filter in the feedback path; this is achieved with the use of a simple first order transfer function element. Figure 3.4 shows the complete speed control loop. Since it is a control system with a first order transfer function and an integrator, the symmetrical optimum has to be applied here, too.

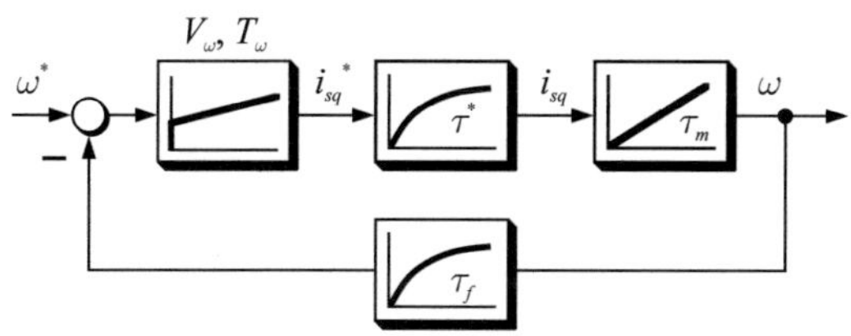

Figure 3.4: Speed control loop with PI controller

3.3 Flux control

Instead of an explicit flux control loop, a steady reference value for the flux-producing component of the stator current i_{sd} used. By setting ${i_{sd}}^* = 0.33$, it can be assumed that, as a first approximation, the machine is constantly excited with nominal flux.

3.4 Experimental results

Experimental results of the cascaded PI controllers in comparison with GPC control are presented in chapter 6.3.3 on page 60 et seqq. for current control and on page 63 et seqq. for speed control, respectively. Further results are shown in chapter 9.5.5 on page 162 et seqq. where the PI controller is compared to direct model-based predictive control with explicit solution of the optimization problem.

4 Predictive control

The linear PID-controllers used in electric drive technology in the beginning were mostly built with analog operational amplifiers and used the control deviation in order to generate an actuating signal. A controller of this type does not posses any knowledge about the plant itself, knowledge about the plant is only required for the controller design. With the availability of inexpensive microcomputers and digital control techniques in drive technology which were developed therefore, the idea to precalculate the plant's behavior via a mathematic model and to determine optimum values for the actuating variables from these precalculated values was born. It was the birth of *predictive* or *precalculating control*.

The first ideas for predictive control methods have been published in the 1960s by Emeljanov [35]. After a rather quiet period in the next decade many predictive control algorithms, which are fundamental for drive technology, like Direct Torque Control (DTC) [1, 113] or Predictive Current Control [55] were developed in the 1980s. Further publications concerned the control of the armature current of DC drives with the help of line-commutated converters [54, 65]; however, this field of research has lost more and more of its significance. After 1990 further publications about predictive drive control appeared; some of them, like [100], show extensions and improvements of known control methods, while other authors, e. g. [37, 118], published completely new control strategies. But at the end, in most cases, these methods turned out to be only further enhancements or combinations of already published fundamental predictive control strategies. Hence, it makes sense to point out the basic functional and fundamental principles of predictive control strategies first.

As already explained in chapter 1, linear PI controllers in cascaded structure which are used in drive control applications, have some basic disadvantages. With predictive control, it is possible to abandon the cascaded loop since in a predictive controller all control variables can be controlled simultaneously in one single controller.

Figure 4.1 shows the typical structure of a predictive controller; as example, a position control of an electric drive is chosen. The measured state variables, namely the machine current I, the rotating speed ω and the position angle φ are processed simultaneously in a model of the drive. With the help of this model, exact information of the current system state can be obtained, which is then transferred into a block called "prediction and calculation". This functional block can be regarded as the heart of a predictive control system; it determines an optimum value for the actuating variable by comparing the current machine state with the desired behavior, i. e. the reference value. By applying this actuating value to the plant via the actuator, the control loop is closed. The calculation of the optimum value for the actuating variable in the block "prediction and calculation" is done depending on the desired optimum condition, such as minimum current error, minimum current distortion or similar things. An evaluation of the efforts for changing the values of the actuating variables can be done, too.

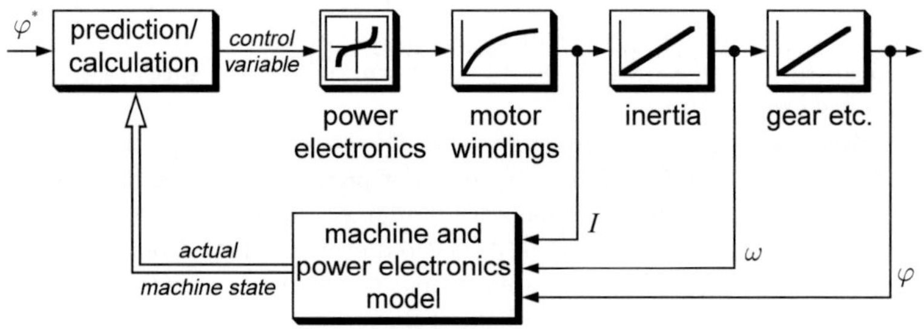

Figure 4.1: Typical structure of a predictive controller

The basic principle described above is common to all predictive control strategies; the only differences are in the functionality of the block "prediction and calculation". Because of these differences in prediction and optimization, predictive algorithms can be classified according to different criteria, namely

- the basic functional principle
- the prediction horizon
- the inverter control

If the predictive controllers published so far are classified according to the above criteria, it can be seen that they can further be classified into different families with typical characteristics.

4.1 Classification based on operational principle

Considering the functional principles of the different predictive control algorithms, it can be seen that these can be classified into three main groups. A decision can be made between *hysteresis-based*, *trajectory-based* and *model-based* strategies. Indeed, these families are not clearly separated from each other and sometimes the transition between them is rather floating.

Hysteresis-based predictive controllers

The basic principle of hysteresis-based control strategies is to keep the value of the controlled variable within a tolerance band or a tolerance area, the so-called hysteresis. The most simple form of such a controller is the well known hysteresis or bang-bang controller. Although in literature bang-bang controllers are not considered to be predictive controllers they do, however, clearly show their typical behavior.

An improved form of a multidimensional bang-bang controller is the predictive current control scheme proposed by Holtz and Stadtfeld [55]. Using this controller, the switching instants are determined by the limits of the tolerance band.

Figure 4.2 shows the functional principle; in this case a circular hysteresis boundary is chosen, whose position in the stator coordinate system is given by the stator current reference vector $\boldsymbol{i}_s{}^*$. If the actual value of the stator current $\boldsymbol{i}_s$ reaches the border of the hysteresis area, the next switching state of the inverter is selected via prediction and optimization: At first, the future trajectories of the stator current vector are precalculated for all possible switching states and the time instant at which the actual current will leave the tolerance band is determined. The basis for these calculations are the well-known mathematical differential equations of electric machines. Besides, it shall be noted that the circular tolerance region itself moves along with the stator current space vector within the complex plane. This movement is indicated with the dotted circle in figure 4.2.

Now the switching state vector which possesses the longest stay-time within the hysteresis is selected. With this optimization criterion, the switching frequency of the inverter is minimized; of course other optimization criteria are also possible, e. g. minimum current distortion or minimum torque ripple.

Hysteresis-based strategies have the advantage that precise knowledge about the system to be controlled is not required. Even with possible model divergences the control error can be kept within the specified limit band by the hysteresis controller. To achieve this, it must always be ensured that the hysteresis controller reacts very quickly if the actual value has gone outside of the hysteresis band. This is a major problem if the hysteresis-based predictive controller is implemented in a digital processor, as the detection of the reference signal crossing the hysteresis band will be done only during the next sampling instant. It may happen that the error has, at this time, already grown to a large value. Hence, hysteresis-based predictive control is more suitable when the realization is done using analog operational amplifiers rather than microcontrollers.

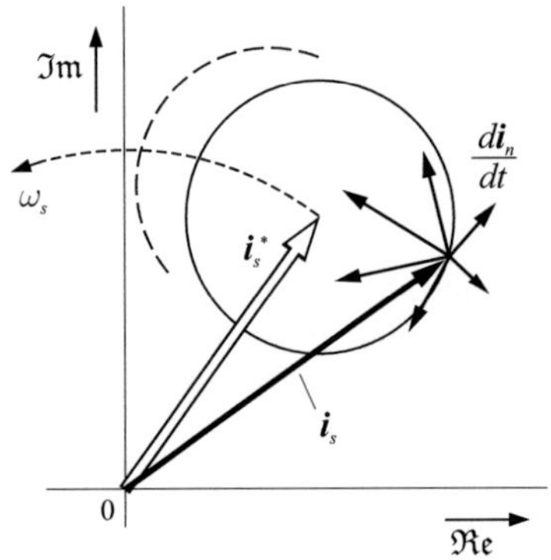

Figure 4.2: Hysteresis-based predictive controller acc. to Holtz/Stadtfeld [55]

Trajectory-based predictive controllers

Trajectory-based control methods are based on the principle to force the system onto precalculated system trajectories. Once the system has been pushed onto one of these trajectories, it remains there because of its own properties until a change is enforced from outside.

The first trajectory-based predictive control scheme has already been published in 1984 by Kennel [65], at that time, however, being a control strategy for a line-commutated thyristor converter. Some time later, well-known control schemes like Direct Self Control (DSC) by Depenbrock [33] or Direct Mean Torque Control (DMTC) by Flach [37] for the control of induction machines were published. Some more schemes, like Sliding Mode Control [35] or Direct Torque Control (DTC) [1,113] are a combination of hysteresis based and trajectory-based schemes, whereas Direct Speed Control (DSPC) by Mutschler [90] can be regarded as a pure trajectory-based control scheme even if some hysteresis based aspects are included in it. In the following, DSPC will be explained as an example of trajectory-based predictive control.

Similar to the schemes of Depenbrock [33] and Takahashi/Noguchi [1,113] the switching states of the inverter are classified into the groups "torque-increasing", "slowly torque-decreasing" and "quickly torque-decreasing". For short time intervals, the inertia of the system as well as the derivatives of the load torque and the machine torque can be assumed to be constant values. Then the system behavior can be represented by a set of parabolas in the speed error/acceleration plane (figure 4.3). These parabolas are also natural trajectories of the system.

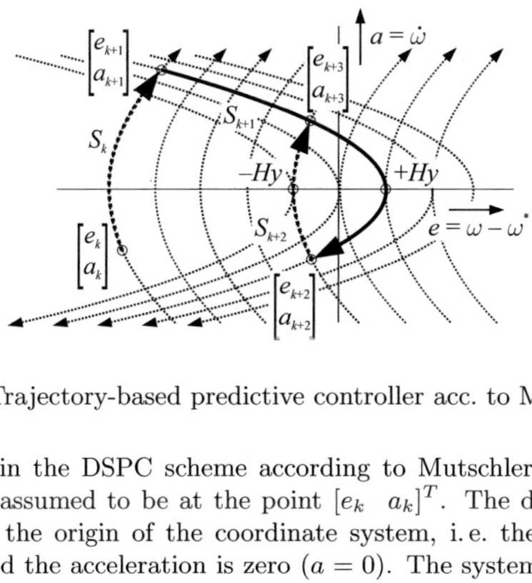

Figure 4.3: Trajectory-based predictive controller acc. to Mutschler [90]

As proposed in the DSPC scheme according to Mutschler [90], the initial system state is assumed to be at the point $[e_k \quad a_k]^T$. The desired operating point is always the origin of the coordinate system, i. e. there is no control error ($e = 0$) and the acceleration is zero ($a = 0$). The system cannot be held in this condition for long time since this would require an infinite switching frequency. Hence, a kind of hysteresis band between $-Hy$ and $+Hy$ is defined; this is described above as the hysteresis-based aspect inside DSPC. In this way, the maximum switching frequency can be reduced to an acceptable value. Apart from that this strategy is purely a trajectory-based scheme. To reach the predefined hysteresis area, at first, a torque increasing switching state S_k is chosen. Now the system state moves along the dotted parabola (trajectory) until $[e_{k+1} \quad a_{k+1}]^T$ is reached. In this point the trajectory of the switching state S_k crosses another parabola for a torque-decreasing switching state S_{k+1}, which will pass through the point $+Hy$. The intersection point between S_k and S_{k+1} in $[e_{k+1} \quad a_{k+1}]^T$ has been determined as the optimum switching instant in advance. Hence, at the correct time instant, switching can take place without any delay and therefore the desired state point $+Hy$ can be reached as fast as possible. Exactly in the precalculated time instant, the inverter is commutated into switching state S_{k+1}. The system state now moves along the new parabola until the point $[e_{k+2} \quad a_{k+2}]^T$ is reached. At this point a torque-increasing switching state S_{k+2} is chosen; the system state now moves along the corresponding trajectory through $-Hy$ until it reaches the parabola of the switching state S_{k+1}, again at the point $[e_{k+3} \quad a_{k+3}]^T$. In steady state operation, the system state keeps moving along the path $+Hy - [e_{k+2} \quad a_{k+2}]^T - -Hy - [e_{k+3} \quad a_{k+3}]^T - +Hy$ and the speed error e is kept within the predefined tolerance band $-Hy$ to $+Hy$.

Trajectory-based predictive control is based on a very precise prediction of the future control system behavior. Hence, in contrast to hysteresis controllers, controllers of this type require an exact model of the system to be controlled. Because of the quite complex precalculation of the system trajectories, these methods are better suited for implementations in the form of digital controllers on microprocessors.

Model-based predictive controllers

Both hysteresis as well as trajectory-based predictive controllers use the current system state to precalculate the value of the controlled variable for the next sampling cycle. The past is not explicitly taken into consideration as it is hidden in the actual system state. Although there is a relationship between hysteresis and trajectory-based predictive control algorithms, model-based strategies (model predictive Control, MPC) are based on completely different ideas. Model-based predictive control methods are able to consider the past and to optimize future values of the actuating variables, not only for the next sampling cycle, but up to a specified future cost or control horizon. Comparing the structure of model-based predictive controllers, it can be seen that they are more like state controllers or Kalman filters rather than the predictive controllers described above.

A more detailed description of model-based predictive controllers is given in chapter 5; thus, it can be omitted at this point.

Family tree

If the predictive control schemes published so far are classified into the tree families mentioned above according to their functional characteristics, a family tree of predictive control strategies can be designed. As shown in figure 4.4 the hysteresis and trajectory-based control schemes are closely related with each other. However, model-based predictive control is based on totally different ideas and thus it forms a class that is independent from the other ones. Since MPC strategies do not differ very much from each other, not all of them are included in the family tree.

4.2 Classification based on prediction horizon and control principle

Another classification method for predictive control algorithms is based on two other criteria. The first distinctive feature is the depth of the precalculation, which is referred to as *prediction horizon*; another partition can be made according to the type of inverter control, the *control principle*: While some predictive controllers immediately calculate optimum inverter switching states, i. e. they control the inverter directly. Other strategies determine a value-continuous control signal which must be synthesized by a modulator before it is passed on to the inverter. Some predictive control methods and their classification in the different families according to the above differentiation criteria are given in table 4.1.

Most of the control schemes which have been investigated in drive technology so far have a prediction horizon of only one single sampling cycle. Well known examples with modulator are e. g. Direct Current Control of Induction Motor Currents by Mayer/Pfaff [87] or Direct Flux Control proposed by Asher et al. [7]. Nevertheless, the biggest part of the one-step predictive controllers belongs to the group of the prediction schemes with direct inverter control, among them such prominent ones like Direct Torque Control [1, 113] and its derivatives as well as Direct Self Control [33] and Direct Speed Control [90].

Predictive control strategies with a prediction horizon of more than one single sampling cycle are exclusively model-based predictive controllers. Thus, they are also referred to as *Long-Range Predictive Control*, abbreviated as *LRPC*. The only scheme of this kind used for drive control so far is Generalized Predictive Control [27, 28]. Its suitability for drive applications has been investigated by Kennel, Linder and Linke in 2001; the results have been published in [66].

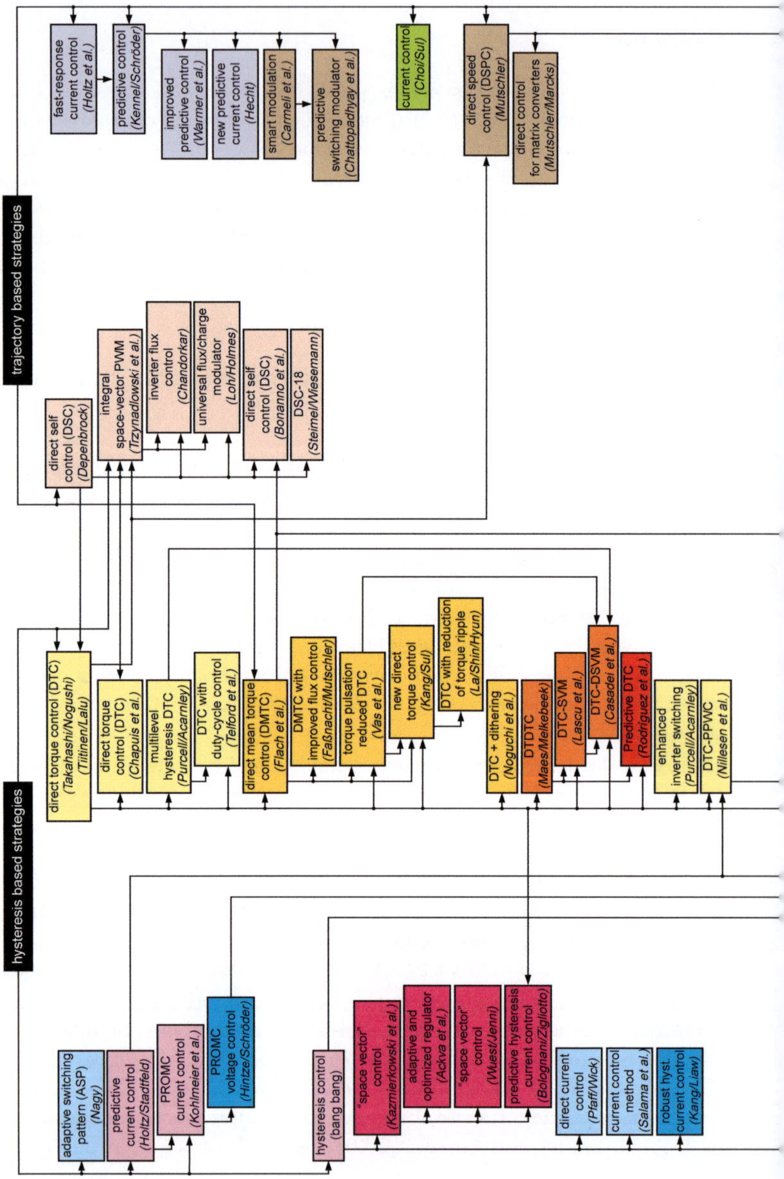

Figure 4.4: Family tree of predictive control schemes

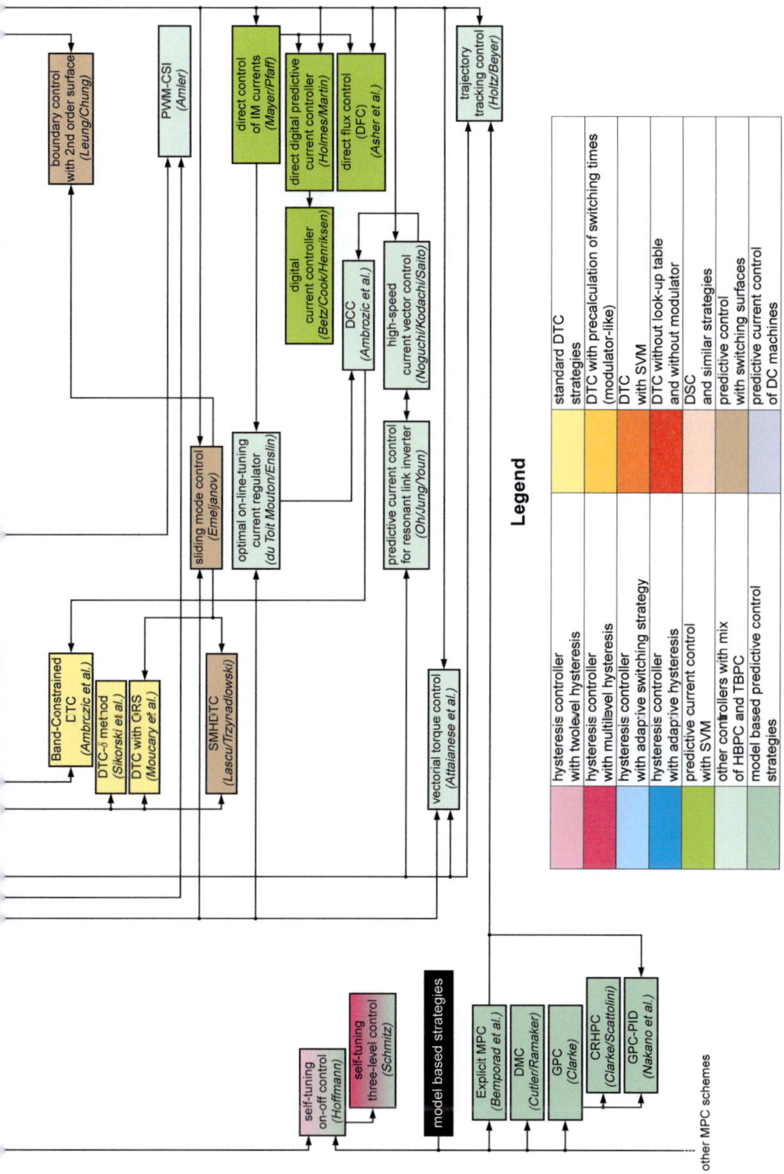

boundary control with 2nd order surface (Leung/Chung)
PWM-CSI (Amler)
direct control of IM currents (Mayer/Pfaff)
direct digital predictive current controller (Holmes/Martin)
direct flux control (DFC) (Asher et al.)
trajectory tracking control (Holtz/Beyer)
digital current controller (Betz/Cook/Henriksen)
DCC (Ambrozic et al.)
high-speed current vector control (Noguchi/Kodachi/Saito)
sliding mode control (Emeljanov)
optimal on-line-tuning current regulator (du Toit Mouton/Enslin)
predictive current control for resonant link inverter (Oh/Jung/Youn)
Band-Constrained DTC (Ambrozic et al.)
DTC-δ method (Sikorski et al.)
DTC with QRS (Moucary et al.)
SMHDTC (Lascu/Trzynadlowski)
vectorial torque control (Attaianese et al.)
self-tuning on-off control (Hoffmann)
self-tuning three-level control (Schmitz)
model based strategies
Explicit MPC (Bemporad et al.)
DMC (Cutler/Ramaker)
GPC (Clarke)
CRHPC (Clarke/Scattolini)
GPC-PID (Nakano et al.)
other MPC schemes
Legend
hysteresis controller with twolevel hysteresis
hysteresis controller with multilevel hysteresis
hysteresis controller with adaptive switching strategy
hysteresis controller with adaptive hysteresis
predictive current control with SVM
other controllers with mix of HBPC and TBPC
model based predictive control strategies
standard DTC strategies
DTC with precalculation of switching times (modulator-like)
DTC with SVM
DTC without look-up table and without modulator
DSC and similar strategies
predictive control with switching surfaces
predictive current control of DC machines

		control principle	
		with modulator	direct
prediction horizon	1	Direct Control of IM Currents Direct Flux Control	Direct Torque Control Direct Self Control Direct Speed Control
	> 1	Generalized Predictive Control	Direct Model Predictive Control

Table 4.1: Classification of predictive control algorithms

Predictive control schemes with a prediction horizon of more than one sampling cycle and with direct inverter control have not been used for drive control so far. This can, to some extent, be justified by the fact that the whole topic of hybrid systems, which is closely related to these control strategies, is a quite new field of research. Chapter 9 deals with the application of such a direct model-based predictive control to an electric drive.

5 Model-based predictive control

The principle of *model-based predictive control* or *model predictive control*—abbreviated *MPC*—was introduced for industrial control applications in the 1970s after first ideas of this strategy have already been published in the 1960s. Subsequently, MPC gained importance mainly in the field of chemical industry and later on it also gained more regard in the academic area. MPC does not denote a special control algorithm, but rather a whole family of controller types. Common characteristic of all these controllers is the principle to determine an optimum value for the actuating variable by using an explicit model of the system to be controlled and by minimizing a cost function.

For a first introduction into the subject of MPC, there exist several papers from technical journals which give an introduction or also a survey about the different model-based control schemes. The article from Morari/Lee [88] is recommended here in which an overview about past, present and future improvement possibilities of MPC are presented. To present the ideas of MPC in an easily understandable manner, the article mainly avoids mathematical equations. For control engineers interested in the mathematical background of MPC, the detailed tutorial by Rawlings [102] is a good reference.

To the family of model-based predictive control belong, among other methods, schemes like Dynamic Matrix Control (DMC), Model Algorithmic Control (MAC), Extended Horizon Adaptive Control (EHAC) and Extended Predictive Self-Adaptive Control (EPSAC) [30]. In contrast to these control strategies, Internal Model Control (IMC, see chapter 10.1) does *not* belong to the MPC class, even though its name suggests this and even if some authors see this in such a way, e. g. García, Prett and Morari [40]. Seborg, on the other hand, classifies it correctly, not as a predictive, but as an "alternative scheme" [109]. A detailed comparison of different MPC strategies is presented by de Keyser et al. [68] and García et al. [40].

As already mentioned in chapter 4, control structures belonging to the family of model-based predictive control have totally different structures than the predictive controllers commonly used in drive technology. Of course, these controllers possess a similar structure, as they also use an explicit and separately identifiable model of the controlled system for the precalculation of the system behavior and therefore also for the selection of optimum values for the actuating variables. But in contrast to conventional predictive controllers used in drive control applications which precalculate the plant's behavior only for the next sampling cycle, MPC controllers consider the future system behavior for more than one sampling cycle [67]. Main advantages of these controllers are:

- Multivariable structures are easily representable.
- System constraints can be handled systematically and can be considered in the model.
- Filtering of measured variables without phase displacement can be integrated without further effort.
- Automatic identification of model parameters is possible.

5.1 Functional principle

The functional principle of an MPC controller can be explained with the help of the structure shown in figure 5.1. Its central part is the model which is used to predict the future behavior of the system to be controlled. The prediction consists of two components:

The free response shows the expected behavior of the system output $\mathbf{y}(t+j)$ assuming future values of the actuating variables being equal to zero.

The forced response forms the additional component of the system response based on the precalculated set of future actuating values $\mathbf{u}(t+j)$.

For linear systems, the entire future system behavior, the *total response*, can be determined as the sum of the free and forced response using the superposition principle. This sum is precalculated up to the *prediction horizon* N_p. According to the prediction horizon, a set of future reference values the system output should be equal to, does also exist. The difference between future reference and precalculated actual values delivers the future control error. An optimization algorithm determines a set of optimum future actuating values $\mathbf{u}(t+j)$ from the expected error, taking system restrictions and the cost function into account. A simple open-loop control scheme would only apply this precalculated future sequence of values for the actuating variables to the system. By using past values of the output and of the actuating variables up to the *past horizon*, this method changes into closed-loop control. Only the first element of the calculated state vector – $\mathbf{u}(t)$ – is applied to the system and afterwards the whole procedure of prediction, optimization and controlling is again repeated for each sampling cycle. Hence, the prediction horizon is shifted forward; this principle is called *Receding Horizon Control* or *RHC*.

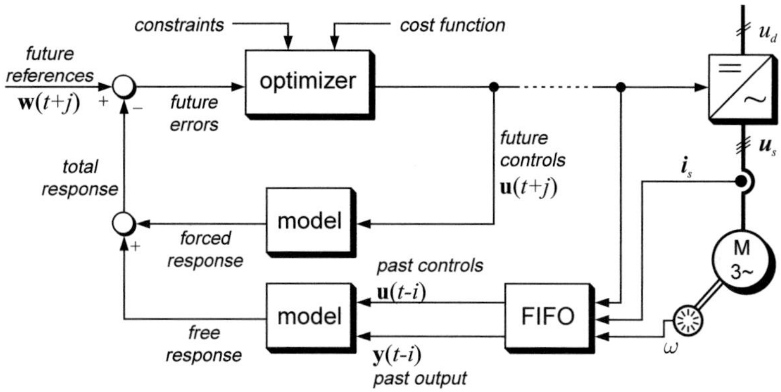

Figure 5.1: Typical structure of an MPC controller

The functional principle of model-based predictive control based on Receding Horizon Control represents a kind of "natural" predictive control, as it is very close to human behavior. For example when driving a car, the driver does not look immediately in front of his car, but he looks far ahead and changes the actuating variables, e. g. the position of the steering wheel, gas pedal and brake before he approaches for instance a red traffic light or a curve. Besides, he precalculates the behavior of the car for a certain distance in front of him up to a finite horizon taking future values of the actuating variables into account, he optimizes the amount of acceleration or braking according to his optimization criteria for this distance and acts accordingly. Like in real MPC, different optimization criteria are possible, leading to different results. If the driver desires the shortest possible duration of his trip, he will accelerate and brake more rapidly than if a reduction of fuel consumption is an optimization criterion.

Due to the precalculation of the system behavior up to the prediction horizon, MPC inevitably leads to a high calculation demand. The calculation complexity can be significantly reduced with the introduction of a so-called *control horizon* N_u. After N_u steps, it is assumed that the steady state is reached and so the controller output remains constant. Figure 5.2 illustrates this situation.

In spite of this modification in the control scheme, still a higher mathematical effort is necessary for model-based control of a system compared to other control algorithms. Hence, main areas of applications for MPC algorithms are, among others, the chemical and process industry [19, 40], since the processes to be controlled there are very well suited for MPC. The time constants of the whole system are rather large (in the range of minutes or even higher). Hence, calculation time is not a problem in this case.

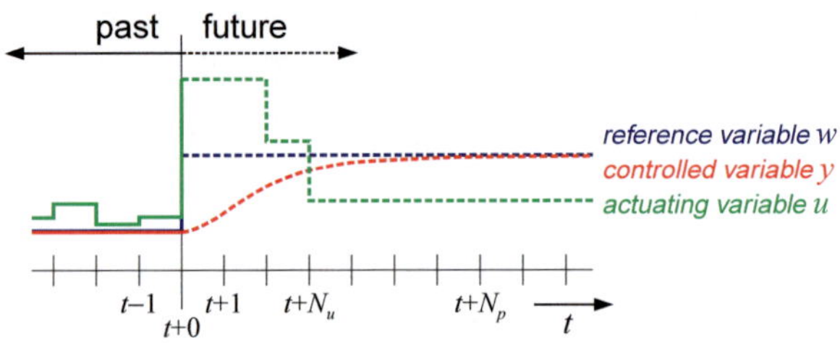

Figure 5.2: Definition of the control and prediction horizon

In the area of electrical drives, however, much higher sampling rates are needed. Several proposals to overcome this problem have been made [11, 18]; nevertheless, publications on practical applications of MPC strategies for drive control are not noted except for the already mentioned investigations by Kennel/Linder/Linke [66].

5.2 Models

Besides the cost function, the model which precalculates the future behavior is one of the crucial points of an MPC scheme. Most systems to be controlled are continuous-time systems so that the description of the system should be made in the time or Laplace domain. In contrast to the controlled systems, the controllers, generally realized with the help of digital computers, are often discrete-time controllers. For the controller design, it is in most cases sufficient to replace it by a continuous-time model, parameterize it and afterwards realize it as a discrete-time digital controller. Nevertheless, it is more advantageous to consider the discrete character of the controller already in the first approach of the controller design and hence to use a discrete-time system description. This also implies that the controlled system is described in the discrete-time domain.

A Laplace transformed transfer function cannot be used now, otherwise the discretization is lost with it. Of course, a representation of the system in the time domain is always possible, however, it has to be made sure that the time t can only assume values $t = kT_0$ with $k \in \mathbb{N}$ and T_0 representing the sampling rate.

The desired representation can be obtained if the discrete-time function $f(kT_0)$ sampled with T_0 is Laplace transformed and if then the term $e^{T_0 s}$ in the Laplace transformed function is replaced with z. In this way, the *Z-transformed* representation $F(z)$ of the discrete-time function $f(kT_0)$ can be obtained. Mathematically the *Z-transformation* can be expressed with the following equation

$$F(z) = \mathfrak{Z}\left\{f(kT_0)\right\} = \sum_{k=0}^{\infty} f(kT_0) z^{-k} \qquad \text{with } z = e^{T_0 s} \tag{5.1}$$

As a result, for $F(z)$ a polynomial in z can be obtained in which the factor z^{-k} means a shift of k sampling cycles. The factor z^{-1} is also called *shift operator* or *backward shift operator.*

From equation (5.1) follows that a Z-transformation *always* results in an infinite series. However, for many functions, closed expressions for their Z-transformed can be given using the characteristics of power series.

For the derivation of a system model in Z-domain, it is not useful to apply the definition of the Z-transformation according to (5.1), as this will lead to very complex mathematical expressions. Since there are transformation tables for the typical structures given in the relevant control system handbooks and textbooks, e. g. [58], the result can normally be obtained much faster by using these tables and by applying the mathematical rules for Z-transformed functions.

5.2.1 State space model

The most simple way to get to a discrete-time representation of a system is the well-known *state space model.* A linear system can easily be represented in a so-called state space description:

$$\frac{d}{dt}\mathbf{x}(t) = \mathbf{A}\mathbf{x}(t) + \mathbf{B}\mathbf{u}(t) \tag{5.2}$$

$$\mathbf{y}(t) = \mathbf{C}\mathbf{x}(t) + \mathbf{D}\mathbf{u}(t) \tag{5.3}$$

In the above equations $\mathbf{A}$ is called the *state matrix*, $\mathbf{B}$ the *input matrix*, $\mathbf{C}$ the *output matrix* and $\mathbf{D}$ the *feedforward matrix* [108], whereas the vectors

$\mathbf{x}(t)$, $\mathbf{u}(t)$ and $\mathbf{y}(t)$ are representing the *state, input* and *output vector*. The equations (5.2) and (5.3) are called *state equation* and *output equation*, respectively [84].

As the model-based control schemes are based on a discrete-time system model, the above equations have to be transformed into a discrete-time representation[1]:

$$\mathbf{x}(k+1) = \mathbf{A_d}\mathbf{x}(k) + \mathbf{B_d}\mathbf{u}(k) \tag{5.4}$$

$$\mathbf{y}(k) = \mathbf{C_d}\mathbf{x}(k) + \mathbf{D_d}\mathbf{u}(k) \tag{5.5}$$

Several strategies to transform the system equations from continuous-time into discrete-time representation will be discussed in chapter 7.

5.2.2 Linear transfer function-based models

For time variable or unknown system parameters an adaptive system model is necessary, whose model parameters can be estimated in a relatively simple way online and which can be updated in real time. State space-based system models are less suitable for such purposes; therefore it is recommendable to use a model which is based on the transfer function of the system to be controlled. For this intended objective, the following model structures have turned out to be especially advantageous. For simplicity all the models in this chapter are presented as SISO models. Nevertheless, an extension to multidimensional MIMO systems is easily possible; a suitable approach is explained in chapter 8 for the CARIMA model.

A survey about the different linear discrete-time models with which the development of an adaptive MPC controller is possible is given in the following. This overview is taken from the book by Kanjilal [62, chapter 2.4].

AR

The *AutoRegressive model*, abbreviated *AR*, can be expressed as follows:

$$y(t) + a_1 y(t-1) + a_2 y(t-2) + \cdots + a_{na} y(t-na) = \xi(t)$$

or briefly

$$A(z^{-1})y(t) = \xi(t)$$

[1] For an easier differentiation between matrices in continuous-time representation and matrices in discrete-time form the matrices in discrete-time representation are marked with the index d.

in which

$$A(z^{-1}) = 1 + a_1 z^{-1} + a_2 z^{-2} + \cdots + a_{na} z^{-na}$$

The function $\xi(t)$ represents the noise or disturbance which is affecting the system. If the noise term cannot be expressed deterministically, white or colored noise is often used for $\xi(t)$.

IAR

The *Integrated AutoRegressive model* or *IAR* is an extension of the AR model with an integrated noise term:

$$A(z^{-1})y(t) = \frac{\xi(t)}{\Delta}$$

in which

$$\Delta = 1 - z^{-1}$$

therefore

$$\Delta y(t) + a_1 \Delta y(t-1) + a_2 \Delta y(t-2) + \cdots + a_{na} \Delta y(t-na) = \xi(t)$$

ARMA

With the *ARMA* or *AutoRegressive Moving Average model*, the noise is represented with an extended term:

$$A(z^{-1})y(t) = C(z^{-1})\xi(t)$$

in which

$$C(z^{-1}) = 1 + c_1 z^{-1} + c_2 z^{-2} + \cdots + c_{nc} z^{-nc}$$

In detail this can be written as:

$$\begin{aligned} y(t) + a_1 y(t-1) + a_2 y(t-2) + \cdots + a_{na} y(t-na) \\ = \xi(t) + c_1 \xi(t-1) + c_2 \xi(t-2) + \cdots + c_{nc} \xi(t-nc) \end{aligned}$$

ARIMA

The abbreviation *ARIMA* stands for *AutoRegressive Integrated Moving Average model.* Similar to the difference between AR and IAR models the noise is, in this case, considered with an integrated structure again:

$$A(z^{-1})y(t) = C(z^{-1})\frac{\xi(t)}{\Delta}$$

or

$$\begin{aligned}\Delta y(t) + a_1\Delta y(t-1) + a_2\Delta y(t-2) + \cdots + a_{na}\Delta y(t-na)\\ = \xi(t) + c_1\xi(t-1) + c_2\xi(t-2) + \cdots + c_{nc}\xi(t-nc)\end{aligned}$$

ARMAX / CARMA

The *ARMAX* or *AutoRegressive Moving Average model with eXogenous inputs* is very similar to the ARMA model, but possesses additional input variables which represent the actuating variables from outside having an effect on the system:

$$A(z^{-1})y(t) = B(z^{-1})u(t-d) + C(z^{-1})\xi(t)$$

in which

$$B(z^{-1}) = b_0 + b_1z^{-1} + b_2z^{-2} + \cdots + b_{nb}z^{-nb}$$

The value d represents the time delay or dead time between the input u and the output y of the system. In systems relevant for drive applications, usually $d = 1$ can be set. Alternatively, additional delays can be absorbed into the polynomial $B(z^{-1})$.

ARMAX models are also known as *CARMA* or *Controlled AutoRegressive Moving Average models.*

ARIMAX / CARIMA

Like the extension of ARMA to ARMAX, ARIMA models can also be extended to *ARIMAX* models, i. e. *AutoRegressive Integrated Moving Average models with eXogenous inputs* by adding additional input variables. The result can be described with the equation:

$$A(z^{-1})y(t) = B(z^{-1})u(t-d) + C(z^{-1})\frac{\xi(t)}{\Delta}$$

Models of the *CARIMA* or *Controlled AutoRegressive Integrated Moving average model* type have an identical structure. They are used e. g. for GPC controllers.

5.2.3 Nonlinear models

Although the area of nonlinear control is still very new and unexplored, a large number of models which represent the behavior of a nonlinear control system exist. However, when a closer look is taken at some of these nonlinear models, they turn out to be linear models using a linearized image of the system to be controlled. As these models are not really nonlinear models in a narrower sense they are not further discussed here. Among the remaining strategies, the *Hammerstein model* and the *Wiener model* are the most prominent ones. Besides these, there is another interesting proposal for modeling nonlinear systems with the so-called *NARMAX model*.

NARMAX

The *Nonlinear AutoRegressive Moving Average model with eXogenous inputs* is described in detail by Leontaritis/Billings [80, 81]. Its main principle is a description of the controlled system similar to the linear ARMAX model, but extended to be suitable for nonlinear, multivariable, discrete-time control systems which can be deterministic as well as stochastic. Basis of a NARMAX model is the following system description for a general multivariable, discrete-time and time-invariant system:

$$\begin{aligned} \mathbf{x}(t+1) &= \mathbf{g}\left[\mathbf{x}(t), \mathbf{u}(t)\right] \\ \mathbf{y}(t) &= \mathbf{h}\left[\mathbf{x}(t), \mathbf{u}(t)\right] \end{aligned}$$

Simplistically, a linear system is considered at first. For such a discrete-time, time-invariant and linear MIMO system the following can be defined:

$$\begin{aligned} \mathbf{x}(t+1) &= \mathbf{g}\left[\mathbf{x}(t), \mathbf{u}(t)\right] = \mathbf{A_d}\mathbf{x}(t) + \mathbf{B_d}\mathbf{u}(t) \\ \mathbf{y}(t) &= \mathbf{h}\left[\mathbf{x}(t), \mathbf{u}(t)\right] = \mathbf{C_d}\mathbf{x}(t) + \mathbf{D_d}\mathbf{u}(t) \end{aligned}$$

This system description in state space, consisting of the matrixes $\mathbf{A_d}$, $\mathbf{B_d}$, $\mathbf{C_d}$ and $\mathbf{D_d}$, can be transformed with the help of a numerical solution technique into a multi-structural linear transfer function model. The obtained description is similar to models which are known from linear control engineering. Indeed, the classical representation forms cannot be generalized for nonlinear systems, which is the reason why the new multi-structural model is necessary.

With the help of some rules of modern analysis and differential geometry, the above approach is transferred and applied to the nonlinear system description. By performing this generalization a recursive, nonlinear and multi-structural I/O model is obtained, which is completely equivalent to the linear description. However, these recursive nonlinear models are only valid in a limited operation range around the equilibrium point.

An enlarged system description is obtained by including stochastic disturbances into the model. As this precisely represents a nonlinear counterpart to the ARMAX model, it is also called NARMAX model.

Hammerstein model

In contrast to most of the linear and nonlinear models explained above, the *Hammerstein model* is not a classical MPC model, but it is also used for other nonlinear controls. Its basic idea is to compose a nonlinear model from a static, nonlinear part followed by a dynamic linear system as shown in figure 5.3. The nonlinearity is expressed e. g. by a polynomial like:

$$\mathbf{r}(t) = f_N\big(\mathbf{u}(t)\big) = \mathbf{\Gamma}_1\mathbf{u}(t) + \mathbf{\Gamma}_2\mathbf{u}^2(t) + \cdots + \mathbf{\Gamma}_{n\alpha}\mathbf{u}^{n\mathbf{\Gamma}}(t)$$

For the linear part, different models are possible. In most cases an ARMAX model (see chapter 5.2.2) is used in which the disturbance term can be neglected.

The parameters of the linear part of the model are usually determined with well-known online parameter identification techniques. The nonlinear system part is static; consequently, no online parameter estimation is performed.

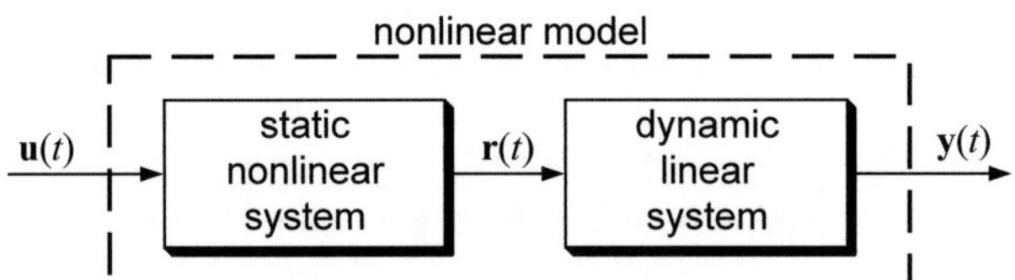

Figure 5.3: Structure of a Hammerstein model

Wiener model

The Wiener model has no relation to the Austrian capital, it is named after the mathematician and physicist Norbert Wiener (1894–1964). It differs from the Hammerstein model by the fact that the nonlinear system is modeled with a dynamic linear system followed by a static nonlinear system (figure 5.4). Like the Hammerstein model the parameters of the linear part are estimated online, whereas the nonlinear part is static.

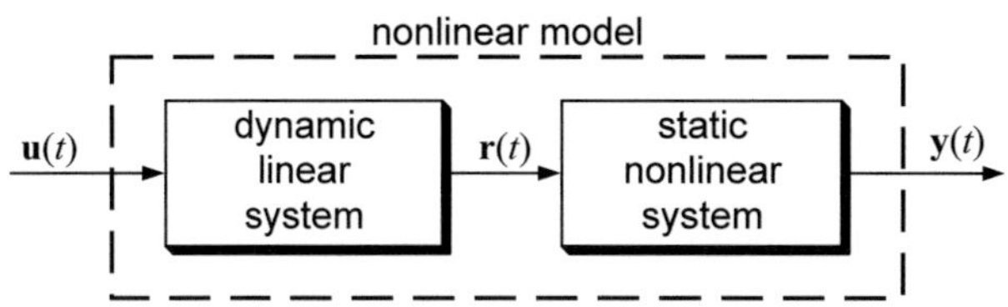

Figure 5.4: Structure of a Wiener model

After in this chapter the general functional principle of model-based predictive controllers as well as different system models suitable for these controllers were introduced, now an MPC controller should be realized in practice. In the following chapters different model-based control strategies are presented and in addition they are investigated experimentally.

6 Generalized Predictive Control

The control strategy *Generalized Predictive Control*—abbreviated *GPC*—also belongs to the group of model-based predictive controllers. It was introduced by Clarke at the University of Oxford in 1987 [27, 28] and makes use of a transfer function-based CARIMA model. The optimization problem is solved analytically by setting the derivative of the cost function equal to zero; thus, the use of mathematically complex solution algorithms like quadratic (QP) or linear programming (LP) is not required.

6.1 "Classical GPC"

In its most simple form, a GPC controller represents a linear SISO controller, i. e. the system to be controlled has only one input and one output variable. Consideration of disturbances or other disturbing signals is not taken into account.

6.1.1 Mathematical derivation

The CARIMA model

Basis of an MPC scheme is always a model of the system or plant which is to be controlled. In a GPC controller a so called CARIMA model is used which is based on the transfer function of the plant. Considering discrete-time structures, this transfer function can commonly be described as a fraction of two polynomials:

$$G(z^{-1}) = \frac{B(z^{-1})}{A(z^{-1})} = \frac{b_0 + b_1 z^{-1} + b_2 z^{-2} + \cdots + b_{nb} z^{-nb}}{1 + a_1 z^{-1} + a_2 z^{-2} + \cdots + a_{na} z^{-na}} \tag{6.1}$$

The above structure can be explained easier if a representation similar to one used in communications engineering for recursive filters (IIR filters) is used. Figure 6.1 shows the equivalent block diagram for equation (6.1). If the system is affected with a dead time, the first elements of the polynomial $B(z^{-1})$ are equal to zero. The equation of the CARIMA model can be derived from (6.1) (see also chapter 5.2.2).

$$A(z^{-1})y(t) = B(z^{-1})u(t-1) + C(z^{-1})\frac{\xi(t)}{\Delta} \tag{6.2}$$

Thereby the last term in (6.2) represents the effect of disturbances. If $\xi(t)$ is white noise, the polynomial can be set to $C(z^{-1}) = 1$. Thus, the equation can be simplified to:

$$A(z^{-1})y(t) = B(z^{-1})u(t-1) + \frac{\xi(t)}{\Delta} \tag{6.3}$$

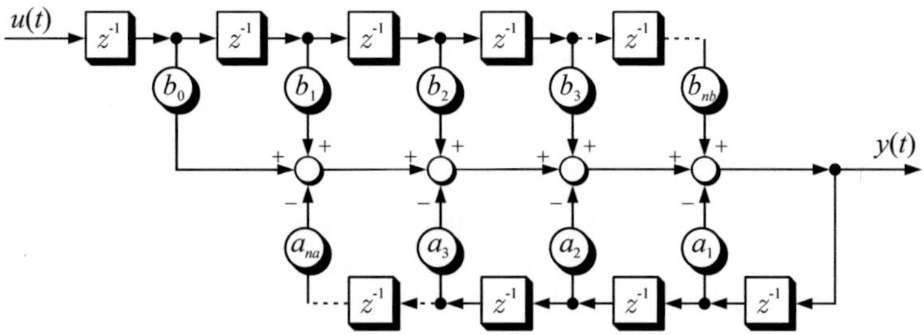

Figure 6.1: Discrete-time transfer function

The j-step ahead predictor

In order to obtain a predictor, the following Diophantine equation[1] is considered:

$$1 = E_j(z^{-1})A(z^{-1})\Delta + z^{-j}F_j(z^{-1})$$

With the help of the definition $\tilde{A}(z^{-1}) = \Delta A(z^{-1})$, the following Diophantine equation can be obtained:

$$1 = E_j(z^{-1})\tilde{A}(z^{-1}) + z^{-j}F_j(z^{-1}) \tag{6.4}$$

The idea behind equation (6.4) is that the reciprocal $1/\tilde{A}(z^{-1})$ is calculated until the remainder can be factorized as $z^{-j}F_j(z^{-1})$. The quotient of the

[1] Refer to appendix A for a description of Diophantine equations.

division is the polynomial $E_j(z^{-j})$. Thus, $E_j(z^{-1})$ is of degree $j-1$ which leads to the fact that $E_j(z^{-1})$ and $F_j(z^{-1})$ can be defined as:

$$\begin{aligned} E_j(z^{-1}) &= e_{j,0} + e_{j,1}z^{-1} + e_{j,2}z^{-2} + \cdots + e_{j,j-1}z^{-(j-1)} \\ F_j(z^{-1}) &= f_{j,0} + f_{j,1}z^{-1} + f_{j,2}z^{-2} + \cdots + f_{j,na}z^{-na} \end{aligned}$$

Now (6.3) is multiplied with $\Delta E_j(z^{-1})z^j$:

$$\tilde{A}(z^{-1})E_j(z^{-1})y(t+j) = E_j(z^{-1})B(z^{-1})\Delta u(t+j-1) + E_j(z^{-1})\xi(t+j)$$

Solving (6.4) for $E_j(z^{-1})\tilde{A}(z^{-1})$ and using it in the above equation, the following result can be obtained:

$$y(t+j) = E_j(z^{-1})B(z^{-1})\Delta u(t+j-1) + F_j(z^{-1})y(t) + E_j(z^{-1})\xi(t+j)$$

Since, as already mentioned, the degree of the polynomial $E_j(z^{-1})$ is $j-1$, all noise components are in the future. Hence, the best possible prediction for y is:

$$\hat{y}(t+j) = G_j(z^{-1})\Delta u(t+j-1) + F_j(z^{-1})y(t) \tag{6.5}$$

in which

$$G_j(z^{-1}) = E_j(z^{-1})B(z^{-1})$$

If the prediction is split into single steps, the following equations for the different prediction steps can be obtained:

$$\begin{aligned} \hat{y}(t+1) &= G_1(z^{-1})\Delta u(t) + F_1(z^{-1})y(t) \\ \hat{y}(t+2) &= G_2(z^{-1})\Delta u(t+1) + F_2(z^{-1})y(t) \\ \hat{y}(t+3) &= G_3(z^{-1})\Delta u(t+2) + F_3(z^{-1})y(t) \\ &\vdots \\ \hat{y}(t+N_p) &= G_{N_p}(z^{-1})\Delta u(t+N_p-1) + F_{N_p}(z^{-1})y(t) \end{aligned}$$

Calculation of the actuating variables

The GPC algorithm uses a quadratic cost function in order to minimize values of the expected control deviations $\hat{y}(t+j) - w(t+j)$ which were predicted with (6.5). At the same time the costs for changing the values of the actuating variables $\Delta u(t+j-1)$ are evaluated, too.

$$J(N_1, N_2, N_u) = \sum_{j=N_1}^{N_2} \mu_j \left(\hat{y}(t+j) - w(t+j)\right)^2 + \sum_{j=1}^{N_u} \lambda_j \left(\Delta u(t+j-1)\right)^2 \quad (6.6)$$

Some MPC methods do not consider the second term which takes the change of the values of the actuating variables into account; other methods evaluate the values of the actuating variables themselves instead of their changes. However, in GPC the entire expression is evaluated. Thereby, the following parameters play a role:

- The parameters N_1 and N_2 are called *lower* and *upper cost horizon*. They indicate the horizon for prediction and optimization. Hence, control deviations in the time interval $t \ldots (t+(N_1-1))$ are not considered in the evaluation. This does e. g. make sense if the plant is affected with a dead time during which the actuating variable has no effect on the controlled variable. In this case, the evaluation of control deviations being in the dead time phase does not make sense because these cannot be corrected in any way. Thus, in order to save computation time, N_1 can be set equal to the value of the dead time.

- The parameter N_u denotes the so called *control horizon*. It does not necessarily need to be equal to the prediction horizon; then, in this case, the optimization algorithm assumes that the value of the actuating variable does not change anymore after N_u steps, i. e. $\Delta u(t+j-1) = 0$ for $j > N_u$.

- The coefficients μ_j weigh the predicted future quadratic control deviations $\hat{y}(t+j) - w(t+j)$. By using an appropriate sequence of μ_j, it is possible to valuate later control deviations higher than control deviations closer to t. Furthermore, by replacing the quadratic error with the absolute value and by selecting $\mu_j \sim j$, an optimum function according to the ITAE criterion can be given.

- The coefficients λ_j weigh the quadratic value of the change of the values of the actuating variables $\Delta u(t+j-1)$. Consequently, in the same way as for weighing of the control deviations, it is possible to valuate values of the actuating variables being not so close in the future, more or less than the ones being close to t.

All these coefficients can be considered as setting parameters with whose help the behavior of the MPC controller can be changed a lot in order to adjust the controller optimally for achieving the desired system behavior. In this way, the behavior of nearly every other predictive and non-predictive controller type can be imposed upon an MPC controller. For a standard GPC controller, Clarke/Mohtadi/Tuffs use the values $\mu_j = 1$ and $\lambda_j = \lambda$ for all j [27]. For reasons of simplicity, also the horizons are often set to $N_1 = 1$ and $N_2 = N_p$.

For the optimization of the values of the actuating variables, a differentiation between values of $u(t)$ being in the future and values being in the past has to be made, since in causal systems values of the actuating variables that are in the past cannot be changed anymore. Thus, the term $G_j(z^{-1})\Delta u(t+j-1)$ in equation (6.5) is splitted into sub-terms concerning the future and the past. The sum of the past output values $F_j(z^{-1})y(t)$ and of the system response to the past values of the actuating variables forms the *free response* $\mathbf{f}$. The rest, consisting of the system response to future values of the actuating variables, is the *forced response*.

$$\mathbf{y} = \mathbf{G}\tilde{\mathbf{u}} + \underbrace{\mathbf{F}(z^{-1})y(t) + \mathbf{G}'(z^{-1})\Delta u(t-1)}_{=\mathbf{f}}$$

$$\mathbf{y} = \underbrace{\mathbf{G}\tilde{\mathbf{u}}}_{\substack{\text{forced}\\ \text{response}}} + \underbrace{\mathbf{f}}_{\substack{\text{free}\\ \text{response}}} \tag{6.7}$$

in which $\tilde{\mathbf{u}} = \Delta\mathbf{u}$. With help of (6.7) and assuming that $\mu_j = 1$ and $\lambda_j = \lambda$ for all j, the cost function (6.6) can be simplified to

$$J = (\mathbf{G}\tilde{\mathbf{u}} + \mathbf{f} - \mathbf{w})^T(\mathbf{G}\tilde{\mathbf{u}} + \mathbf{f} - \mathbf{w}) + \lambda\tilde{\mathbf{u}}^T\tilde{\mathbf{u}}$$

In order to minimize this equation, it is first expanded into separate terms, whereby

$$J = \tilde{\mathbf{u}}^T(\mathbf{G}^T\mathbf{G} + \lambda\mathbf{I})\tilde{\mathbf{u}} + \tilde{\mathbf{u}}^T\mathbf{G}^T(\mathbf{f} - \mathbf{w}) + (\mathbf{f} - \mathbf{w})^T\mathbf{G}\tilde{\mathbf{u}} + (\mathbf{f} - \mathbf{w})^T(\mathbf{f} - \mathbf{w})$$

results. Because of the identity,

$$\tilde{\mathbf{u}}^T\mathbf{G}^T(\mathbf{f} - \mathbf{w}) = (\mathbf{f} - \mathbf{w})^T\mathbf{G}\tilde{\mathbf{u}}$$

J can be further simplified to

$$J = \tilde{\mathbf{u}}^T(\mathbf{G}^T\mathbf{G} + \lambda\mathbf{I})\tilde{\mathbf{u}} + 2\tilde{\mathbf{u}}^T\mathbf{G}^T(\mathbf{f} - \mathbf{w}) + (\mathbf{f} - \mathbf{w})^T(\mathbf{f} - \mathbf{w}) \tag{6.8}$$

The minimum of this equation is determined by setting the derivative of J equal to zero:

$$\frac{\partial J}{\partial \tilde{\mathbf{u}}} = 2(\mathbf{G}^T\mathbf{G} + \lambda\mathbf{I})\tilde{\mathbf{u}} + 2\mathbf{G}^T(\mathbf{f} - \mathbf{w}) \stackrel{!}{=} 0$$

Solving the equation for $\tilde{\mathbf{u}}$, the equation

$$\tilde{\mathbf{u}} = (\mathbf{G}^T\mathbf{G} + \lambda\mathbf{I})^{-1} \cdot \mathbf{G}^T(\mathbf{w} - \mathbf{f}) \tag{6.9}$$

for optimum future values of the actuating variables can be obtained. Since only the first element of $\tilde{\mathbf{u}}$, which is actually $\Delta u(t)$, is required, the law for the value of the actuating variable that has to be applied to the plant can be derived from equation (6.9):

$$u(t) = u(t-1) + \tilde{\mathbf{g}}^T(\mathbf{w} - \mathbf{f}) \tag{6.10}$$

Thereby $\tilde{\mathbf{g}}^T$ is the first row of matrix $(\mathbf{G}^T\mathbf{G} + \lambda\mathbf{I})^{-1} \cdot \mathbf{G}^T$.

Recursion of the Diophantine equation

The Diophantine equation (6.4) has to be solved in order to calculate $E_j(z^{-1})$ and $F_j(z^{-1})$. Although other GPC algorithms which do *not* require the calculation of Diophantine equations are available, e. g. the ones proposed by Albertos and Ortega [3], in the following the "classical" solution will be presented. For this purpose, the Diophantine equations for the steps j and $j+1$ are arranged first and then they are subtracted from each other:

$$\begin{array}{c|l} - & 1 = E_j(z^{-1})\tilde{A}(z^{-1}) + z^{-j}F_j(z^{-1}) \\ + & 1 = E_{j+1}(z^{-1})\tilde{A}(z^{-1}) + z^{-(j+1)}F_{j+1}(z^{-1}) \\ \hline & 0 = \left(E_{j+1}(z^{-1}) - E_j(z^{-1})\right)\tilde{A}(z^{-1}) + z^{-j}\left(z^{-1}F_{j+1}(z^{-1}) - F_j(z^{-1})\right) \end{array}$$

The term $E_{j+1}(z^{-1}) - E_j(z^{-1})$ is of degree j, since $E_j(z^{-1})$—as already explained above—is of degree $j-1$. Thus, one can write:

$$E_{j+1}(z^{-1}) - E_j(z^{-1}) = \tilde{R}(z^{-1}) + r_j z^{-j} \tag{6.11}$$

in which $\tilde{R}(z^{-1})$ is a polynomial of degree $\leqq j-1$ and r_j is a real number. This leads to the Diophantine equation

$$0 = \tilde{R}(z^{-1})\tilde{A}(z^{-1}) + z^{-j}\left(z^{-1}F_{j+1}(z^{-1}) - F_j(z^{-1}) + r_j\tilde{A}(z^{-1})\right)$$

From $\tilde{a}_0 = 1$ follows that $\tilde{R}(z^{-1}) = 0$ because of the shifting of the remaining terms with z^{-j}. As a result, the equation is simplified to

$$z^{-1}F_{j+1}(z^{-1}) = F_j(z^{-1}) - r_j\tilde{A}(z^{-1})$$

Because of $\tilde{a}_0 = 1$, the following expressions result by comparing the coefficients of the above polynomial equation:

$$f_{j,0} = r_j \tag{6.12}$$

$$f_{j+1,i} = f_{j,i+1} - r_j \tilde{a}_{i+1} \tag{6.13}$$

From (6.11), (6.12) and the abovementioned fact that $\tilde{R}(z^{-1}) = 0$, the recursion equation for the polynomials $E_j(z^{-1})$ can be obtained.

$$E_{j+1}(z^{-1}) = E_j(z^{-1}) + f_{j,0} z^{-j} \tag{6.14}$$

The final terms of the recursion equations can be obtained from (6.4) for $j = 1$ and under consideration of $\tilde{a}_0 = 1$:

$$E_1(z^{-1}) = e_0 = 1 \tag{6.15}$$

$$F_1(z^{-1}) = z\big(1 - \tilde{A}(z^{-1})\big) \tag{6.16}$$

From (6.14) it can be seen that all coefficients of the polynomial $E_{j+1}(z^{-1})$ from 0 to $-(j-1)$ are identical to the ones of the polynomial $E_j(z^{-1})$, i.e. both polynomials differ only in the term $f_{j,0}z^{-j}$, which is newly added to the polynomial $E_{j+1}(z^{-1})$. Hence, the single polynomials $E_j(z^{-1})$ can be written as:

$$\begin{aligned}
E_1(z^{-1}) &= e_0 \\
E_2(z^{-1}) &= e_0 + e_1 z^{-1} \\
E_3(z^{-1}) &= e_0 + e_1 z^{-1} + e_2 z^{-2} \\
&\vdots \\
E_j(z^{-1}) &= e_0 + e_1 z^{-1} + e_2 z^{-2} + \cdots + e_{j-1} z^{-(j-1)} \\
E_{j+1}(z^{-1}) &= e_0 + e_1 z^{-1} + e_2 z^{-2} + \cdots + e_{j-1} z^{-(j-1)} + e_j z^{-j} \\
&= E_j(z^{-1}) + e_j z^{-j}
\end{aligned}$$

in which

$$e_j = f_{j,0}$$

Calculation of the free and forced response

The prediction equation (6.5) is split into the forced and into the free response (6.7). A calculation of the polynomials $G_j(z^{-1})$ is unnecessary if it is easier to calculate the matrix $\mathbf{G}$ and the vector $\mathbf{f}$ individually.

First the matrix $\mathbf{G}$ is determined. The individual rows of this matrix can be calculated from the expression $G_j(z^{-1}) = E_j(z^{-1})B(z^{-1})$:

$$\begin{aligned} G_j(z^{-1}) &= E_j(z^{-1})B(z^{-1}) \\ &= \left(e_0 + e_1 z^{-1} + e_2 z^{-2} + \cdots + e_{j-1} z^{-(j-1)}\right) \\ &\quad \cdot \left(b_0 + b_1 z^{-1} + b_2 z^{-2} + \cdots + b_{nb} z^{-nb}\right) \end{aligned}$$

The coefficients of the polynomial $G_j(z^{-1})$, which was derived according to the above expression, are then inserted into the jth row of the matrix $\mathbf{G}$. Since this matrix should be used for the calculation of the forced response, no terms $u(t+j)$ with $j < 0$ shall enter the equation, because these values would be in the past and thus, they would not be relevant for the forced, but for the free response. Therefore, the polynomial $G_j(z^{-1})$ may only be taken into consideration up to the degree $j-1$. As it can be seen in the upper expression, in this case, the matrix elements $g_{j,0} \ldots g_{j,j-1}$ and $g_{j+1,0} \ldots g_{j+1,j-1}$ are all identical, i. e. the matrix rows differ only in the newly added element $g_{j+1,j}$. Therefore:

$$\mathbf{G} = \begin{bmatrix} g_0 & 0 & 0 & \cdots & 0 \\ g_1 & g_0 & 0 & \cdots & 0 \\ g_2 & g_1 & g_0 & \cdots & 0 \\ \vdots & \vdots & \vdots & \ddots & \vdots \\ g_{N_p-1} & g_{N_p-2} & g_{N_p-3} & \cdots & g_0 \end{bmatrix}$$

with

$$g_i = e_0 b_i + e_1 b_{i-1} + \cdots + e_i b_0$$

If $i > nb$ all summations with $b_k, k > nb$ are void.

Now the following method is adopted for the calculation of the free response: The old $\mathbf{y}$- and $\mathbf{u}$-values are combined to a new vector $\mathbf{yu}$. Accordingly, a matrix $\mathbf{FG}'$ is composed from the coefficients of the polynomial matrices $\mathbf{F}(z^{-1})$ and $\mathbf{G}'(z^{-1})$ which results in:

$$\mathbf{f} = \mathbf{F}(z^{-1})y(t) + \mathbf{G}'(z^{-1})\Delta u(t-1) = \mathbf{FG}' \cdot \mathbf{yu}$$

in which

$$\mathbf{F}(z^{-1}) = \begin{bmatrix} F_1(z^{-1}) \\ F_2(z^{-1}) \\ \vdots \\ F_{N_p}(z^{-1}) \end{bmatrix} = \begin{bmatrix} f_{1,0} + f_{1,1}z^{-1} + \cdots + f_{1,na}z^{-na} \\ f_{2,0} + f_{2,1}z^{-1} + \cdots + f_{2,na}z^{-na} \\ \vdots \\ f_{N_p,0} + f_{N_p,1}z^{-1} + \cdots + f_{N_p,na}z^{-na} \end{bmatrix}$$

$$\mathbf{G}'(z^{-1}) = \begin{bmatrix} g'_{1,0} + g'_{1,1}z^{-1} + \cdots + g'_{1,nb-1}z^{-(nb-1)} \\ g'_{2,0} + g'_{2,1}z^{-1} + \cdots + g'_{2,nb-1}z^{-(nb-1)} \\ \vdots \\ g'_{N_p,0} + g'_{N_p,1}z^{-1} + \cdots + g'_{N_p,nb-1}z^{-(nb-1)} \end{bmatrix}$$

$$\mathbf{FG}' = \begin{bmatrix} f_{1,0} & f_{1,1} & \cdots & f_{1,na} & g'_{1,0} & g'_{1,1} & \cdots & g'_{1,nb-1} \\ f_{2,0} & f_{2,1} & \cdots & f_{2,na} & g'_{2,0} & g'_{2,1} & \cdots & g'_{2,nb-1} \\ \vdots & \vdots & \vdots & \vdots & \vdots & \vdots & \vdots & \vdots \\ f_{N_p,0} & f_{N_p,1} & \cdots & f_{N_p,na} & g'_{N_p,0} & g'_{N_p,1} & \cdots & g'_{N_p,nb-1} \end{bmatrix}$$

$$\mathbf{yu} = \begin{bmatrix} y(t) \\ y(t-1) \\ \cdots \\ y(t-na) \\ \Delta u(t-1) \\ \Delta u(t-2) \\ \cdots \\ \Delta u(t-nb) \end{bmatrix}$$

The individual elements $g'_{j,i}$ are calculated in the same way as the elements of $\mathbf{G}$ with the multiplication of $E_j(z^{-1})$ and $B(z^{-1})$. However, in this case, only the values of $u(t+j)$ that are in the past are of interest. Hence, the expression for the calculation is:

$$g'_{j,i} = \sum_{k=0}^{j-1} e_k b_{j+i-k} \qquad j = 1 \ldots N_p;\ i = 0 \ldots (nb-1)$$

As summands with $b_k, k > nb$ can also not exist in this case, all terms with $j+i-k > nb$ must be omitted. Because of the degree of $E_j(z^{-1})$, the summation is only valid for $k = 0 \ldots (j-1)$, since for higher values of k the terms e_k do not exist anymore.

Now the calculated free response $\mathbf{f}$ can be used in equation (6.10) and the value of the actuating variable $u(t)$ for the next sampling cycle can be determined.

6.1.2 Experimental results

Since practical experiments showed that a GPC controller without stabilization through a design polynomial $T(z^{-1})$ cannot be practically used because of the highly increased sensitivity related to model deviations or noise influences, there are no measurement results at this point.

6.2 GPC with filter

If the measured controlled variable contains harmonics and harmonic-free sampling cannot be guaranteed, then the harmonics on the measuring signal lead to a disturbed actuating variable. Especially in very fast acting controllers like MPC methods, this leads to strong oscillations in the actuating variable, since the controller tries to counteract against the apparent control deviation. A similar problem appears if there are errors in the model. In this case, there are two workarounds. Either the controller can be made more sluggish so that it does not react to higher frequencies in the control deviation anymore, or a low-pass filter is inserted into the feedback path. For the latter case, however, it shall be kept in mind that the low-pass filter, if not considered in the controller design, can lead to an unstable behavior of the entire system, since in dynamic operation the feedback value of the controlled variable is no more equal to the real value of the controlled variable. Hence, if the fed back values of the controlled variable are smoothed via a filter, the controller has to be adapted correspondingly which in reality means to make it slower. Considering extremely large filter time constants, this leads to the fact that not the actual plant itself, but the low-pass filter is controlled, whereby the dynamics of the overall system suffer to a great extent.

GPC controllers offer the possibility to integrate a filter into the controller itself. The low-pass filter behavior is considered in the controller's internal model, so that a reduced sensitivity of the controller for high frequency harmonics results *without influencing the overall dynamics.* In this case, the GPC model actually behaves like a Kalman filter [72], which is also well known for its delay-free filtering properties.

For a better understanding of GPC with internal filter, the MPC structure from figure 5.1 is modified according to figure 6.2. As already explained, the complete system response is precalculated from the actuating variable $u(t)$, the controlled variable $y(t)$ and this response is used in the control law for calculating optimum future values for the actuating variables under consideration of the reference value $w(t)$. In addition, figure 6.2 shows the possibility to use a parameter adaptive plant model, but this shall not be discussed further.

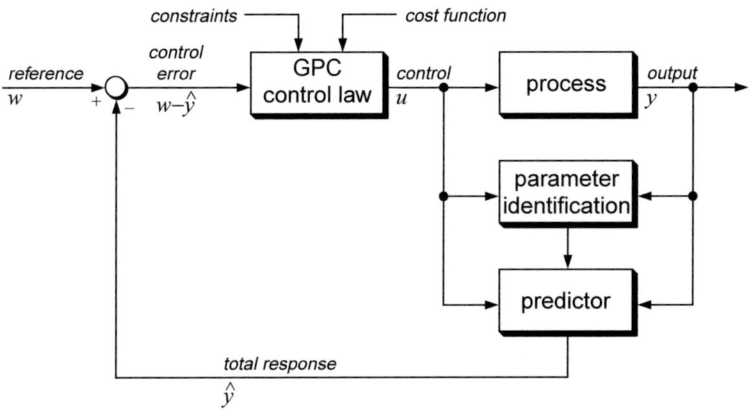

Figure 6.2: Structure of a GPC controller without filter

If the measuring signals containing high-frequency noise should be filtered for the suppression of the unwanted, high-frequency disturbances, the controller structure has to be modified by inserting two low-pass filters with identical time constants as shown in figure 6.3. The measured values of the actual controlled variable $y(t)$ as well as the values of the actuating variable $u(t)$ are filtered in the same way. Then the filtered values $u^f(t)$ and $y^f(t)$ are used for parameter estimation and prediction of the complete response. With marginal changes in the structure of the GPC control law, in the parameter identification and in the predictor in the GPC controller, filtering can be taken into account, which allows a—related to the entire system—delay-free filtering of unwanted harmonics.

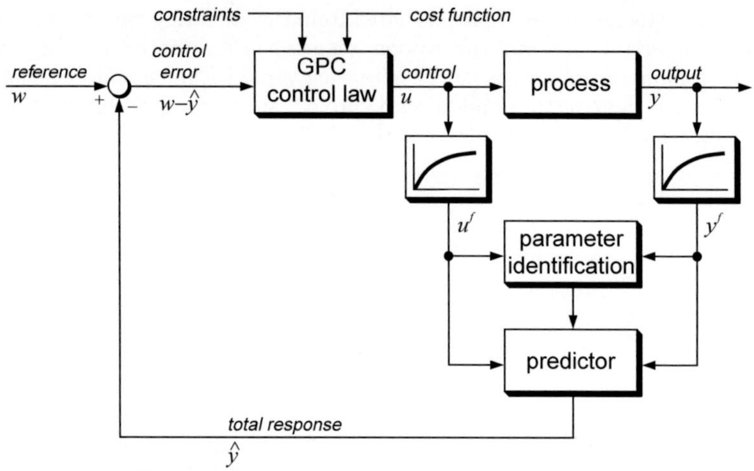

Figure 6.3: Structure of a GPC controller with filter

6.2.1 Mathematical derivation

The CARIMA model

The actual CARIMA model described in chapter 6.1.1 does in principle not change if a low-pass filter is inserted. However, now it cannot be assumed that $\xi(t)$ in equation (6.2) is white noise. Analog to this, the polynomial $C(z^{-1})$ cannot be set equal to 1, as now the disturbances should explicitly be taken into consideration. However, as the disturbances are normally not exactly known, $C(z^{-1})$ is replaced by a *design polynomial* $T(z^{-1})$. This is selected in such a way that it acts like a low-pass filter. Therefore, equation (6.3) can be modified to

$$A(z^{-1})y(t) = B(z^{-1})u(t-1) + T(z^{-1})\frac{\xi(t)}{\Delta} \tag{6.17}$$

with

$$T(z^{-1}) = t_0 + t_1 z^{-1} + t_2 z^{-2} + \cdots + t_{nt} z^{-nt}$$

in which for a filter normally $t_0 = 1$ is set.

The j-step ahead predictor

Taking the design polynomial into consideration, the Diophantine equation, which is already known from equation (6.4), changes to

$$T(z^{-1}) = E_j(z^{-1})\tilde{A}(z^{-1}) + z^{-j}F_j(z^{-1}) \tag{6.18}$$

With the help of the same transformations as already described in chapter 6.1.1 on page 40 et seq., the following equation can be obtained from the above equation and (6.17):

$$\begin{aligned} T(z^{-1})y(t+j) = &\, E_j(z^{-1})B(z^{-1})\Delta u(t+j-1) \\ &+ F_j(z^{-1})y(t) \\ &+ T(z^{-1})E_j(z^{-1})\xi(t+j) \end{aligned}$$

Since here $E_j(z^{-1})$ is of degree $j-1$, too, all noise components are in the future and thus, they cannot appear in the precalculation. Hence, the best possible prediction of y is given by

$$T(z^{-1})\hat{y}(t+j) = G_j(z^{-1})\Delta u(t+j-1) + F_j(z^{-1})y(t)$$

in which again

$$G_j(z^{-1}) = E_j(z^{-1})B(z^{-1})$$

Now, Camacho/Bordons [20] propose the derivation of an optimum prediction equation for the calculation of $\hat{y}(t+j)$ directly from the Δu- and y-values via a direct approach with a Diophantine equation. However, Clarke et al. [28] use the values $y^f(t)$ and $u^f(t)$, which are filtered with $1/T(z^{-1})$, instead of using $y(t)$ and $u(t)$. Considering this approach, the prediction equation for $\hat{y}(t+j)$ results to

$$\hat{y}(t+j) = G_j(z^{-1})\Delta u^f(t+j-1) + F_j(z^{-1})y^f(t)$$

Since future values of the actuating variables are logically *not* filtered with $T(z^{-1})$, $G_j(z^{-1})$ can be separated in the following way:

$$G_j(z^{-1}) = G_j{}'(z^{-1})T(z^{-1}) + z^{-j}\Gamma_j(z^{-1}) \tag{6.19}$$

For the case that $T(z^{-1}) = 1$ (no filtering), the equation $G_j(z^{-1}) = G_j{}'(z^{-1})$ is valid. Now the prediction is separated into the free and forced response by equation (6.19). For this reason the prediction equation results into:

$$\hat{y}(t+j) = \underbrace{G_j{}'(z^{-1})\Delta u(t+j-1)}_{\text{forced response}} + \underbrace{F_j(z^{-1})y^f(t) + \Gamma_j(z^{-1})\Delta u^f(t-1)}_{\text{free response}} \tag{6.20}$$

Calculation of the actuating variables

The optimization itself is done in the same way as for standard GPC without filter (chapter 6.1.1 on page 42 et seqq.). Equation (6.20) can be summarized to

$$\mathbf{y} = \mathbf{G}'\tilde{\mathbf{u}} + \mathbf{f}'$$

With the help of the cost function (6.6), the equation for the precalculation of optimum future values for the actuating variables can be obtained:

$$\tilde{\mathbf{u}} = (\mathbf{G'}^T\mathbf{G}' + \lambda\mathbf{I})^{-1} \cdot \mathbf{G'}^T(\mathbf{w} - \mathbf{f}') \tag{6.21}$$

Analog to standard GPC, it can be stated that $\mathbf{G}' \neq \begin{bmatrix} G_1' & G_2' & G_3' & \cdots \end{bmatrix}^T$ because $\mathbf{G}'$ only considers the future parts of $\tilde{\mathbf{u}}$ for the forced response. Also in this case, only the first element of $\tilde{\mathbf{u}}$, namely $\Delta u(t)$, is required for the real controller. Hence, the value of the actuating variable for the next sampling cycle results from equation (6.21) :

$$u(t) = u(t-1) + \tilde{\mathbf{g}}'^T(\mathbf{w} - \mathbf{f}') \tag{6.22}$$

in which $\tilde{\mathbf{g}}'^T$ is the first row of the matrix $(\mathbf{G'}^T\mathbf{G}' + \lambda\mathbf{I})^{-1} \cdot \mathbf{G'}^T$.

Recursion of the Diophantine equation

Compared to GPC without filter, the Diophantine equation has to be changed because of the consideration of the design polynomial (compare equation (6.4) with (6.18)). Nevertheless the same strategy for the calculation of $E_j(z^{-1})$ and $F_j(z^{-1})$ can be used. If the Diophantine equations for j and $j+1$ are formed and then subtracted from each other as described in chapter 6.1.1 on page 44 et seq., $T(z^{-1})$ can be omitted. Therefore, the recursion equations (6.12) to (6.14) are also valid in this case.

The final terms of the recursion are obtained from equation (6.18) for $j = 1$ which results in:

$$E_1(z^{-1}) = e_0 = \frac{t_0}{\tilde{a}_0} = 1 \qquad \text{because } \tilde{a}_0 = 1 \text{ and } t_0 = 1 \tag{6.23}$$

$$F_1(z^{-1}) = z\big(T(z^{-1}) - \tilde{A}(z^{-1})\big) \tag{6.24}$$

Calculation of the free and forced response

In order to simplify the precalculation of the free response if a filter is used, vectors and matrices are combined, too. In this case, the filtered values $y^f(t)$ and $\Delta u^f(t)$ from the past form the vectors $\mathbf{y}^f$ and $\mathbf{u}^f$ which are then again combined to a new vector $\mathbf{yu}^f$. Instead of the polynomial matrices $\mathbf{F}(z^{-1})$ and $\mathbf{G}'(z^{-1})$ now $\mathbf{F}(z^{-1})$ and $\mathbf{\Gamma}(z^{-1})$ are used. Therefore:

$$\mathbf{f}' = \mathbf{F}(z^{-1})y^f(t) + \mathbf{\Gamma}(z^{-1})\Delta u^f(t-1) = \mathbf{F\Gamma} \cdot \mathbf{yu}^f$$

in which

$$\mathbf{F}(z^{-1}) = \begin{bmatrix} F_1(z^{-1}) \\ F_2(z^{-1}) \\ \vdots \\ F_{N_p}(z^{-1}) \end{bmatrix} = \begin{bmatrix} f_{1,0} + f_{1,1}z^{-1} + \cdots + f_{1,na}z^{-na} \\ f_{2,0} + f_{2,1}z^{-1} + \cdots + f_{2,na}z^{-na} \\ \vdots \\ f_{N_p,0} + f_{N_p,1}z^{-1} + \cdots + f_{N_p,na}z^{-na} \end{bmatrix} \text{(unchanged)}$$

$$\mathbf{\Gamma}(z^{-1}) = \begin{bmatrix} \Gamma_1(z^{-1}) \\ \Gamma_2(z^{-1}) \\ \vdots \\ \Gamma_{N_p}(z^{-1}) \end{bmatrix} = \begin{bmatrix} \gamma_{1,0} + \gamma_{1,1}z^{-1} + \cdots + \gamma_{1,nb-1}z^{-(nb-1)} \\ \gamma_{2,0} + \gamma_{2,1}z^{-1} + \cdots + \gamma_{2,nb-1}z^{-(nb-1)} \\ \vdots \\ \gamma_{N_p,0} + \gamma_{N_p,1}z^{-1} + \cdots + \gamma_{N_p,nb-1}z^{-(nb-1)} \end{bmatrix}$$

$$\mathbf{F\Gamma} = \begin{bmatrix} f_{1,0} & f_{1,1} & \cdots & f_{1,na} & \gamma_{1,0} & \gamma_{1,1} & \cdots & \gamma_{1,nb-1} \\ f_{2,0} & f_{2,1} & \cdots & f_{2,na} & \gamma_{2,0} & \gamma_{2,1} & \cdots & \gamma_{2,nb-1} \\ \vdots & \vdots & \vdots & \vdots & \vdots & \vdots & \vdots & \vdots \\ f_{N_p,0} & f_{N_p,1} & \cdots & f_{N_p,na} & \gamma_{N_p,0} & \gamma_{N_p,1} & \cdots & \gamma_{N_p,nb-1} \end{bmatrix}$$

$$\mathbf{yu}^f = \begin{bmatrix} y^f(t) \\ y^f(t-1) \\ \cdots \\ y^f(t-na) \\ \Delta u^f(t-1) \\ \Delta u^f(t-2) \\ \cdots \\ \Delta u^f(t-nb) \end{bmatrix}$$

As equation (6.20) shows, the prediction consists of a part for the forced and one for the free response. First, in the same way as for the “filterless” case (see chapter 6.1.1 on page 46 et seqq.) the following approach is made:

$$\begin{aligned} G_j(z^{-1}) &= E_j(z^{-1})B(z^{-1}) \\ &= \left(e_0 + e_1 z^{-1} + e_2 z^{-2} + \cdots + e_{j-1} z^{-(j-1)}\right) \\ &\cdot \left(b_0 + b_1 z^{-1} + b_2 z^{-2} + \cdots + b_{nb} z^{-nb}\right) \end{aligned}$$

Under consideration of equation (6.19)

$$E_j(z^{-1})B(z^{-1}) = G_j{}'(z^{-1})T(z^{-1}) + z^{-j}\Gamma_j(z^{-1}) \qquad (6.25)$$

can be obtained. Since $\mathbf{G}'$ should be used for the prediction of the forced response, terms $u(t+j)$ with $j < 0$ must not appear in $G_j{}'(z^{-1})$. Therefore, in the same way as for $\mathbf{G}$, it can be stated that

1. the matrix elements $g'_{j,0} \dots g'_{j,j-1}$ and $g'_{j+1,0} \dots g'_{j+1,j-1}$ are all identical and
2. the matrix rows only differ in the added element $g'_{j+1,j}$.

Hence, $\mathbf{G}'$ can be defined as:

$$\mathbf{G}' = \begin{bmatrix} g'_0 & 0 & 0 & \cdots & 0 \\ g'_1 & g'_0 & 0 & \cdots & 0 \\ g'_2 & g'_1 & g'_0 & \cdots & 0 \\ \vdots & \vdots & \vdots & \ddots & \vdots \\ g'_{N_p-1} & g'_{N_p-2} & g'_{N_p-3} & \cdots & g'_0 \end{bmatrix}$$

A procedure described by Clarke et al. [28] for the determination of the elements of $\mathbf{F\Gamma}$ is suggested here. Similar to the approach for the calculation of the Diophantine equation for $E_j(z^{-1})$ and $F_j(z^{-1})$, only the equation (6.19) for $G_{j+1}(z^{-1})$ and $G_j(z^{-1})$ is formed and then $G_{j+1}(z^{-1})$ and $G_j(z^{-1})$ are subtracted from each other:

$$\begin{array}{c|rl} + & G_{j+1}(z^{-1}) &= G_{j+1}{}'(z^{-1})T(z^{-1}) + z^{-(j+1)}\Gamma_{j+1}(z^{-1}) \\ - & G_j(z^{-1}) &= G_j{}'(z^{-1})T(z^{-1}) + z^{-j}\Gamma_j(z^{-1}) \\ \hline & e_j z^{-j} B(z^{-1}) &= g'_j z^{-j} T(z^{-1}) + z^{-j}\left(z^{-1}\Gamma_{j+1}(z^{-1}) - \Gamma_j(z^{-1})\right) \end{array}$$

The following expressions can be obtained by comparing the coefficients of the above polynomial equation:

$$\begin{aligned} g'_j &= \frac{e_j b_0 + \gamma_{j,0}}{t_0} \\ \gamma_{j+1,i-1} &= \gamma_{j,i} + e_j b_i - g'_j t_i \end{aligned}$$

As for the last element $\gamma_{j+1,i}$, the summand $\gamma_{j,i+1}$ does of course not exist, in this case the following applies:

$$\gamma_{j+1,i} = e_j b_{i+1} - g'_j t_{i+1}$$

As it can be easily seen from the above definition, the counter variable i varies from 0 to $\max(nb-1, nt-1)$. Since $T(z^{-1})$ in most cases represents a simple low-pass filter, only the elements $t_0 = 1$ and t_1 exist. Normally it can be assumed that $nb \geqq nt$ and therefore $i = 0 \ldots (nb-1)$. As a matter of course, the non-existing terms with $b_k, k > nb$ and $t_k, k > nt$, respectively, are omitted.

The final terms of the recursion can also be obtained by coefficient comparison from equation (6.25) for $j = 1$:

$$g'_0 = \frac{e_0 b_0}{t_0}$$
$$\gamma_{1,i} = e_0 b_{i+1} - g'_0 t_{i+1}$$

Thus, the matrices $\mathbf{G}'$ and $\mathbf{\Gamma}(z^{-1})$ can be calculated, whereas only $\mathbf{\Gamma}(z^{-1})$ is required for the calculation of the free response $\mathbf{f}'$. For the calculation of optimum values for the actuating variables according to equation (6.21), $\mathbf{G}'$ is also not completely necessary since only the first element of $\tilde{\mathbf{u}}$ is used (see page 52). However, the values g'_j are required for determining the elements of the matrix $\mathbf{\Gamma}(z^{-1})$ so that in any case $\mathbf{G}'$ must be completely calculated.

With the help of the recursions and equations shown above, the free response $\mathbf{f}'$ can be determined, be substituted into equation (6.22) and then the next value of the actuating variable $u(t)$ that should be applied to the plant can be calculated.

6.2.2 Simulations

The typical behavior of a GPC controller with filter polynomial $T(z^{-1})$ is verified by the following simulation results. It can be shown that low-pass filtering of the sampled actuating variable can be done without influence on the dynamics of the overall system.

In figure 6.4, the filtering is turned off by setting $T(z^{-1}) = 1$. Since the measured values of the controlled variables are not affected by any disturbances, the control shows ideal behavior. The reference value is reached in three sampling steps.

If the measuring signal is disturbed with a rectangle-shaped noise signal with half the sampling frequency, the signals shown in figure 6.5(a) can be

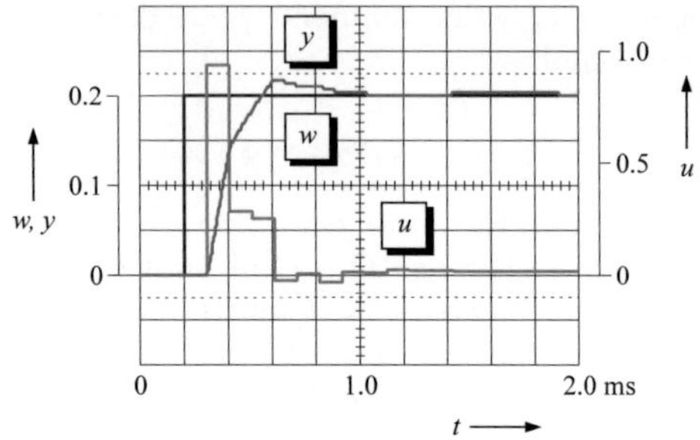

Figure 6.4: Simulation: Reference action without noise disturbances

obtained. Although the amplitude of the noise signal is only 0.005 the controller output signal shows strong oscillating behavior; the amplitude of $u(t)$ is approximately 0.2. The actual value of the controlled variable $y(t)$ of the controlled plant is indeed damped by the plant time constants; despite this fact, the huge periodical steps of the actuating variable between +0.2 and −0.2 are causal for the jerky behavior of the system.

If the internal filter features of a GPC controller are enabled now by choosing the filter polynomial $T(z^{-1}) = 1 - 0.95z^{-1}$, the behavior of the controller can be significantly improved (figure 6.5(b)). Still the same rectangular-shaped disturbance signal is added to the measured values of the controlled variable. Because of the enabled filter, the amplitude of the oscillations of the actuating variable is now only about 0.02, which corresponds to an improvement of the factor 10. Accordingly, the controlled variable shows nearly no more superimposed oscillations. Despite this, the step response of the entire system has not significantly changed; the reference value is still reached in three sampling steps.

6.3 Cascade control with GPC controllers

As a first approach the PI controllers of a conventional field-oriented control in cascaded structure are replaced by GPC controllers that are also cascaded. The signal flow graph of the control is shown in figure 6.6. As it can be seen here, no real flux controller is implemented, but the flux is kept approximately constant because of a constant reference value for the field-producing stator current component i_{sd}.

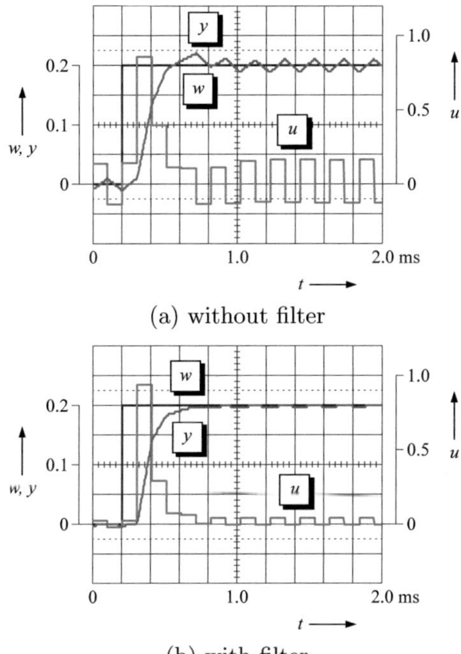

(a) without filter

(b) with filter

Figure 6.5: Simulation: Reference action with noise disturbances

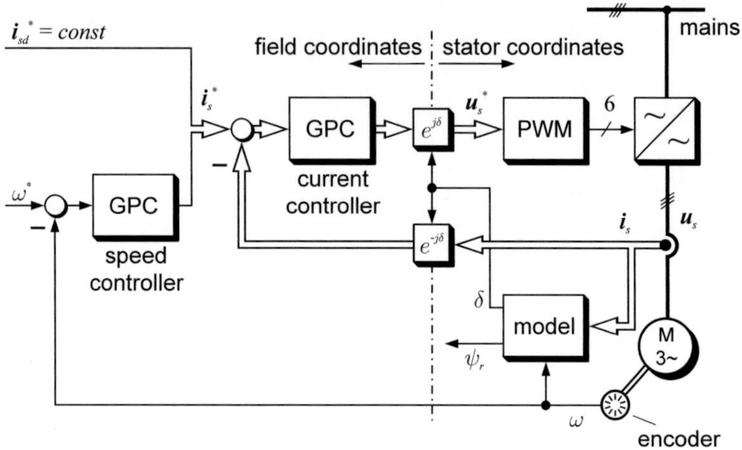

Figure 6.6: Field-oriented drive control with GPC controllers

6.3.1 Current control

Current control with a GPC controller is, in principle, based on the same plant model as the conventional control (figure 3.3 on page 15). Indeed, the discrete-time approach of GPC offers the possibility of considering dead times with a value $n \cdot T_0$ directly in the model by shifting the coefficients of the polynomial $B(z^{-1})$ by z^{-n}. If a CARIMA model according to equation (6.2) or (6.3) is used, a dead time of $1 \cdot T_0$ is already implemented in the model because of the approach with $u(t-1)$ instead of $u(t)$. Therefore, the inverter dead time does not have to be considered for the determination of the CARIMA model parameters since it has already been defined in the model itself. The resulting simplified signal flow graph of the current control loop is shown in figure 6.7.

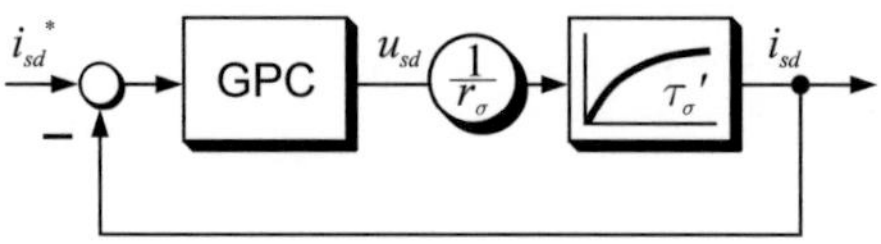

Figure 6.7: Current control loop with GPC controller

Using the standard conversion tables, e. g. from [58], the Z-transfer function of the plant can be obtained:

$$G(z^{-1}) = \frac{\frac{1}{r_\sigma}\left(1 - e^{-\frac{\tau_0}{\tau_\sigma'}}\right) z^{-1}}{1 - e^{-\frac{\tau_0}{\tau_\sigma'}} z^{-1}} = \frac{B(z^{-1})}{A(z^{-1})}$$

As the prediction is normally not done with $A(z^{-1})$, but with $\tilde{A}(z^{-1}) = \Delta A(z^{-1})$, the individual coefficients of $\tilde{A}(z^{-1})$ are calculated. Thus, for $\tilde{A}(z^{-1})$ and $B(z^{-1})$ the following coefficients can be obtained:

$$\tilde{a}_0 = 1 \qquad b_0 = 0$$

$$\tilde{a}_1 = -\left(1 + e^{-\frac{\tau_0}{\tau_\sigma'}}\right) \qquad b_1 = \frac{1}{r_\sigma}\left(1 - e^{-\frac{\tau_0}{\tau_\sigma'}}\right)$$

$$\tilde{a}_2 = e^{-\frac{\tau_0}{\tau_\sigma'}}$$

6.3.2 Speed control

The approach for designing the GPC controller for the speed control loop is in principle the same as described for conventional PI cascade control in chapter 3.2 on page 15. The inner current control loop is replaced with a first order transfer function (PT_1-block) with the time constant τ^*. Figure 6.8 shows the signal flow graph of the GPC speed control loop. As the internal filter characteristics of GPC are used, the additional low-pass filter in the feedback path can be significantly smaller than a corresponding one for PI control.

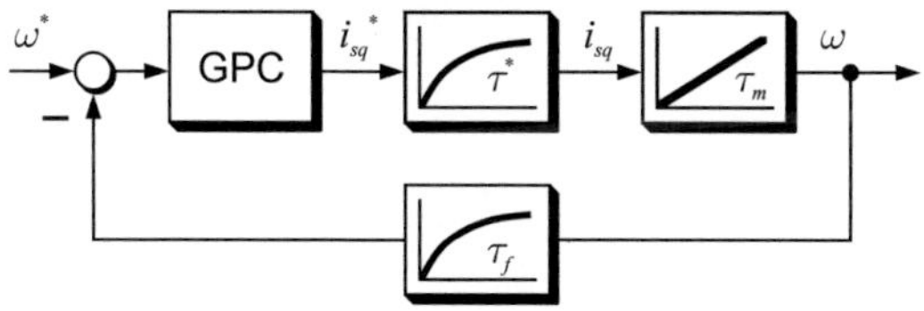

Figure 6.8: Speed control loop with GPC controller

Even the sum of the time constants of both PT_1-blocks is much smaller than the mechanical time constant τ_m, hence, they are not considered in the design of the GPC controller. It can be shown by simulation that including both first order transfer function blocks into the plant model merely increases the order of the system and because of this also the calculation matrices of the GPC controller, but the control performance cannot be improved. Therefore, the model design can be limited to a pure integrator; the discrete-time transfer function then becomes to:

$$G(z^{-1}) = \frac{\frac{\tau_0}{\tau_m} z^{-1}}{1 - z^{-1}}$$

In this case, the coefficients of the polynomials $\tilde{A}(z^{-1})$ and $B(z^{-1})$ are:

$$\tilde{a}_0 = 1 \qquad b_0 = 0$$
$$\tilde{a}_1 = -2 \qquad b_1 = \frac{\tau_0}{\tau_m}$$
$$\tilde{a}_2 = 1$$

6.3.3 Experimental results

Current control

To evaluate the behavior of the GPC controller, it is meaningful to compare it with a PI controller. Because of this, current control of an asynchronous machine was implemented with a conventional PI and with a model-based predictive controller. The PI controller was designed as described in chapter 3.1 on page 13 et seqq. according to the symmetrical optimum method and then it was further optimized empirically. The best possible values that could be obtained are $T_i = 0.33$ and $V_i = 2.3$. The parameters of the GPC controller can be taken from table 6.1.

N_p	N_u	na	nb	nt
4	2	2	1	1

$\tilde{a}_0$	$\tilde{a}_1$	$\tilde{a}_2$	b_0	b_1	t_0	t_1	λ
1.0	-1.9947	0.9947	0.0	0.1650	1.0	-0.95	0.003

Table 6.1: Parameter settings for the GPC current controller

Figure 6.9 shows the large-signal behavior of the closed control loop; therein figure 6.9(a) shows the behavior of a PI controlled drive, while the results in figure 6.9(b) were obtained with a GPC controller. In both cases, the torque-producing current component i_{sq} was step-changed from $i_{sq}{}^* = 0$ to $i_{sq}{}^* = 0.4$. The results show that the GPC controller generates a smaller overshoot than the PI controller. In both cases, the rise time is approximately identical, because it is essentially limited by the available power of the actuator. It can be seen that in this case no significant advantage is offered by the use of a model-based predictive controller.

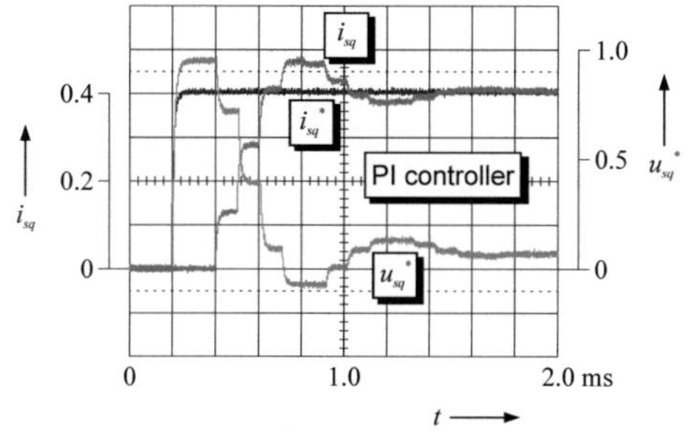

(a) PI controller

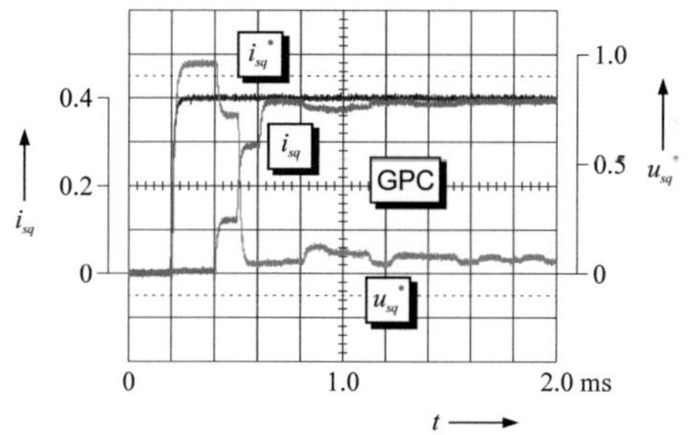

(b) GPC controller

Figure 6.9: Current control: Large-signal behavior

The small-signal behavior of both controllers can be seen in figure 6.10. Comparing the actual values of the output (controlled) variable of the current control loop with the PI controller shown in figure 6.10(a) to the behavior of the GPC controller shown in figure 6.10(b), it can be seen that the PI controller does not make use of the full available actuator power because of the small step-change of the reference value of only $\Delta i_{sq}{}^* = 0.1$, while the model-based predictive controller uses the available actuator power in an optimal way due to its totally different characteristics. Hence, the GPC controller shows superior performance in the small-signal behavior compared to the PI controller.

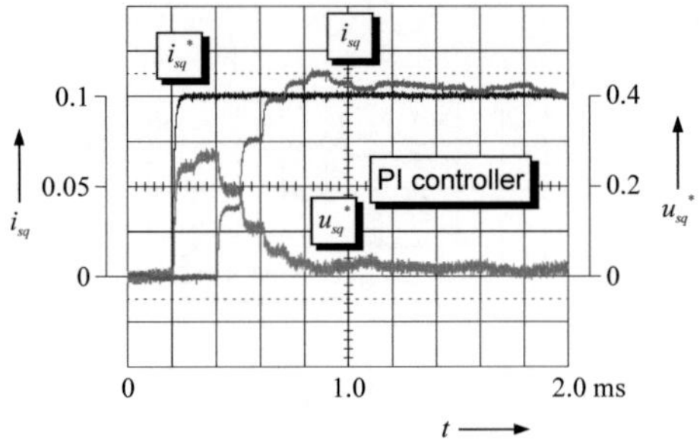

(a) PI controller

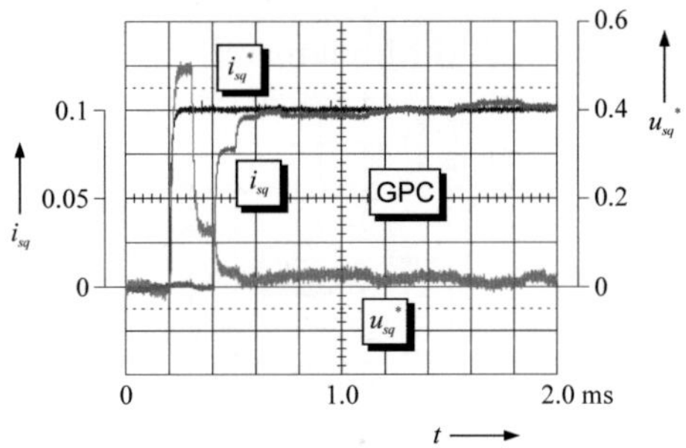

(b) GPC controller

Figure 6.10: Current control: Small-signal behavior

In contrast to a PI controller, an MPC controller with a transfer function-based plant model considers not only present values of the actuating and of the controlled variables, but also considers the past for the calculation of optimum values for the actuating variables. If constant disturbances appear in a control loop, a model-based predictive controller knows them since the past is also considered in the precalculations. In drive control, this case appears typically in the form of the back EMF, which is proportional to the drive speed that acts like a disturbance in the current control loop. Because of the much larger time constants of the speed control loop, the value of this disturbance can be assumed to be constant for the current control.

The measurements shown in figure 6.11 were taken with a normalized rotational speed $\omega = 0.4$ of the machine. As shown in figure 6.11(a), using a simple PI current controller, the rise time is significantly longer because of the back EMF generated by the rotation of the machine. However, the GPC controller (figure 6.11(b)) shows much better behavior, since the disturbing influence of the back EMF is known to the GPC controller from the past. Hence, as a machine is rarely operated in standstill mode and thus, the back EMF does nearly always exist, the MPC controller shows better results than the PI controller in large-signal behavior, too.

Speed control

As already done for current control, GPC speed control should also be compared to speed control with a PI controller. For conventional control, the initial settings obtained according to chapter 3.2 on page 15 were verified experimentally and then optimized. Finally the values $T_\omega = 64.0$ and $V_\omega = 17.2$ have been selected as the best possible controller parameters.

It is well known that the dynamics of the speed control loop are significantly lower than those of the current control loop. However, for technical programming reasons, both current and speed controllers have to be operated with the same sampling frequency. Now the prediction horizon of an MPC controller should approximately match the time constants of the plant to be controlled. Thus, a relatively high prediction horizon of $N_p = 200$ is necessary for a GPC speed controller. Nevertheless, since the control horizon is only $N_u = 1$, the mathematical complexity is limited to a certain extent. A higher value for N_u did not improve the controller behavior so that a larger control horizon obviously makes no sense. The remaining optimal parameters of the GPC controller are given in table 6.2.

The large-signal behavior of a speed control loop gives no or only few information about the quality of the controller because in this case the dynamics are solely limited by the available inverter power. Hence, only the small-signal behavior is examined for the comparison between PI and GPC controller for speed control. The results are presented in figure 6.12.

In drive technology, it has quite often to be dealt with the problem that the measured speed signal must be filtered with a low-pass filter, as otherwise it

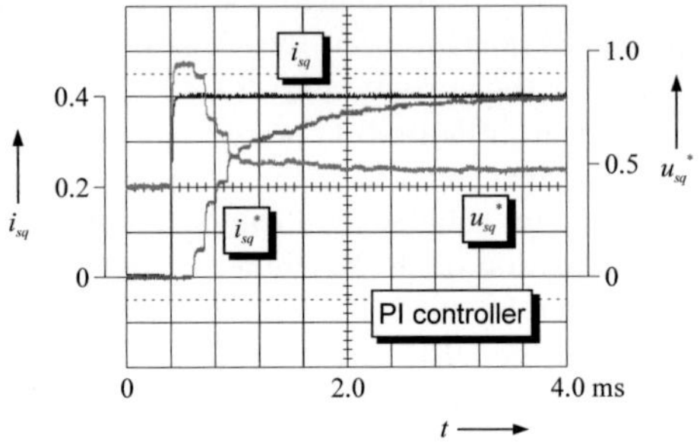

(a) PI controller

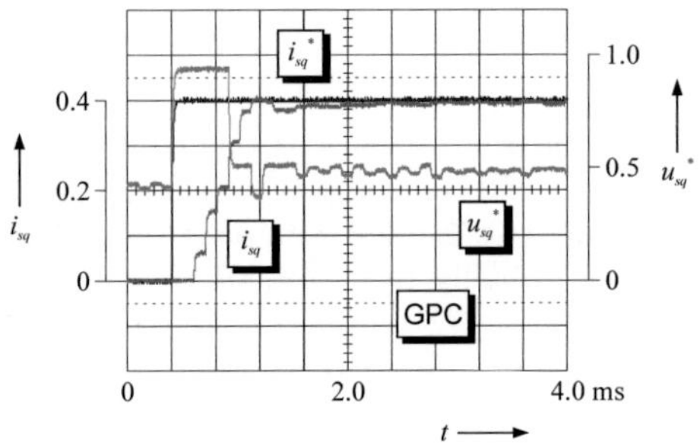

(b) GPC controller

Figure 6.11: Current control: Large-signal behavior at $\omega = 0.4$

N_p	N_u	na	nb	nt
200	1	2	1	1

$\tilde{a}_0$	$\tilde{a}_1$	$\tilde{a}_2$	b_0	b_1	t_0	t_1	λ
1.0	-2.0	1.0	0.0	0.00013	1.0	-0.999	0.1

Table 6.2: Parameter settings for a GPC speed controller

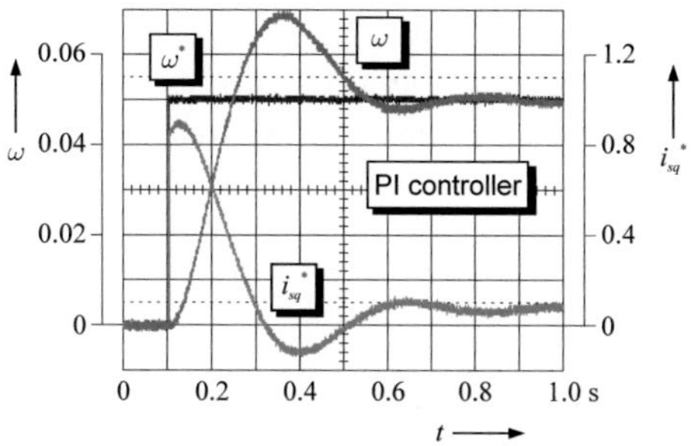

(a) PI controller

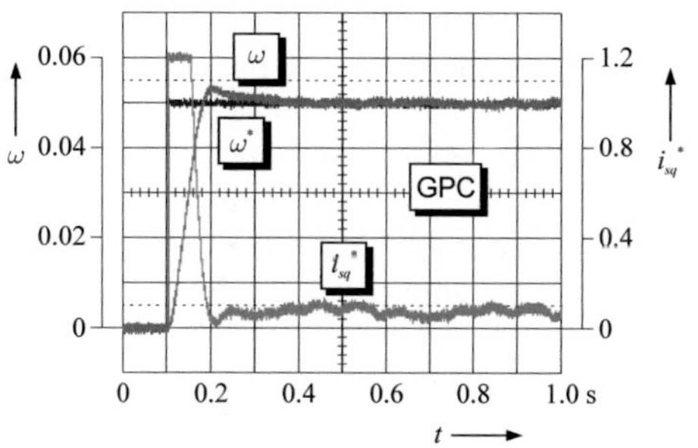

(b) GPC controller

Figure 6.12: Speed control: Small-signal behavior

is not useable for control. This problem appears especially when incremental encoders with low resolution are used, because the speed signal received by counting the increments per sampling cycle changes with distinct jumps. Of course, the problems described above do not appear when tacho generators are used as speed encoders; however, tacho generators are more expensive than incremental encoders, they also have to be maintained regularly and they are not useful at low speeds because of the proportionality between the amplitude of the output signal and the actual speed. This is why they are normally not used in modern drive technology. When a low-pass filter is inserted into the feedback path this has to be considered in the controller design, leading to the fact that the dynamics of the overall control system are deteriorated. However, in the case of using a model-based predictive controller, the filtering of the measured controlled variable can be shifted into the controller itself, so that no negative effect with respect to the overall system dynamics results. This has already been explained in chapter 6.2.

As it can easily be seen the PI controller in figure 6.12(a) shows the typical behavior of a controller designed according to the symmetrical optimum. The rise time is quite high, also due to the fact that the machine is supplied only with a maximum of 90 percent of the nominal current. Altough the overshoot could be reduced by changing the controller parameters V_ω and T_ω, this would lead to even slower dynamics of the overall system. The selection of parameter values resulting into a less sluggish behavior must not be done, since the control will then become unstable because of the low-pass filter in the feedback path.

However, a model-based predictive controller, whose response is shown in figure 6.12(b), can use the advantage of the internal filtering of the measured signal in the speed control loop. The reduced rise and settling time can easily be seen. Here, the dynamics are not limited by the controller itself, only by the maximum value of the torque-producing component of the stator current i_{sq} to $i_{sq,max} = 1.2$. The GPC controller in the speed control loop is clearly superior to conventional PI control because of the significantly shorter rise time and the almost not existing overshoot. Figure 6.13 once more shows the behavior of a GPC speed controller with a zoomed time scale.

6.3.4 Computation times

For the conventional control methods in drive technology it is extremely important to keep the computation time of the control algorithm as short as possible, because otherwise the high sampling rate necessary for good control quality cannot be obtained. However, as the measurement results given in table 6.3

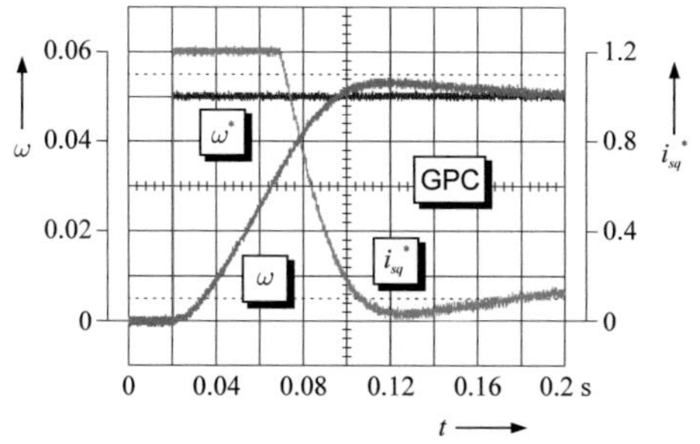

Figure 6.13: Speed control: Small-signal behavior, zoomed time scale

show, this can also be achieved with model-based predictive control based on the GPC principle. A GPC controller in the current control loop does not need more computation time than a PI controller. However, when a model-based predictive controller is used as a speed controller, the high prediction horizon with $N_p = 200$ required here becomes apparent. As a result the computation time is increased to a value being 4 times higher compared to the value of a conventional PI controller. Nevertheless, the total time is still within a range which permits the use of GPC even for fast switching inverters. In addition, if the performance gains represented in chapter 6.3.3 on page 63 et seqq. are considered the additional computational expenses are counterbalanced in many cases.

Type	PI	GPC
Current controller	$2 \times 2.3\,\mu s$	$2 \times 2.3\,\mu s$
Speed controller	$2.3\,\mu s$	$8.8\,\mu s$

Table 6.3: Comparison of computation times for PI and GPC controller

As process computer for these and all other measurements a commercial PC with an AMD Duron® processor (900 MHz clock frequency) was used together with the real time operating system Linux/RTAI. Thus the results presented here are in terms of absolute values not necessarily comparable to results that can be achieved with standard microcomputers that are normally used in drive technology; however, the ratio of the computation times for a PI and a GPC controller should be comparable.

The calculation of the system matrices for the GPC controller involves a matrix inversion. For the drive-technological examples presented in this work, parameter adaptation of the system model was omitted; hence, the matrix inversion can be carried out offline and thus, it does not belong to the time critical parts of the program. However, if an adaption of the model parameters should be implemented, then the inversion of the system matrices has to be done during the runtime of the control process. Although this does not have to be done within one sampling cycle, a time-saving programming of this task should be taken into consideration. A comparison of different procedures for matrix inversion and the computation times required for these tasks can be found in appendix F.

7 Discrete-time machine model for current control

Since a cascaded controller structure has some drawbacks due to its functional principle [67] and since model-based predictive control offers the possibility of MIMO control, this advantage should be utilized. As a first example, a MIMO current controller for field-oriented control of an induction machine should be derived, i. e. a current controller that simultaneously controls the flux-producing as well as the torque-producing component of the stator current $\boldsymbol{i}_s$. For such a multidimensional control, a multidimensional plant model is necessary, too. As a model-based predictive controller should be used, the MIMO plant model of the machine has to be discrete-time.

7.1 Derivation

A linear system can be clearly described in the so-called state space representation (see equations (5.2) and (5.3) on page 31). For a machine model according to this representation, it makes sense to use the following variables for the vectors $\mathbf{x}$, $\mathbf{u}$ and $\mathbf{y}$:

$$\mathbf{x} = \begin{bmatrix} i_{sd} \\ i_{sq} \\ \psi_{rd} \end{bmatrix}, \qquad \mathbf{u} = \begin{bmatrix} u_{sd} \\ u_{sq} \end{bmatrix}, \qquad \mathbf{y} = \begin{bmatrix} i_{sd} \\ i_{sq} \\ \psi_{rd} \end{bmatrix}$$

The following state space representation results from the known machine equations (2.11)–(2.13):

$$\frac{d}{d\tau} \begin{bmatrix} i_{sd} \\ i_{sq} \\ \psi_{rd} \end{bmatrix} = \begin{bmatrix} -\frac{1}{\tau_\sigma{}'} & \omega_s & \frac{k_r}{r_\sigma \tau_r \tau_\sigma{}'} \\ -\omega_s & -\frac{1}{\tau_\sigma{}'} & -\frac{k_r \omega}{r_\sigma \tau_\sigma{}'} \\ \frac{l_h}{\tau_r} & 0 & -\frac{1}{\tau_r} \end{bmatrix} \cdot \begin{bmatrix} i_{sd} \\ i_{sq} \\ \psi_{rd} \end{bmatrix} + \begin{bmatrix} \frac{1}{r_\sigma \tau_\sigma{}'} & 0 \\ 0 & \frac{1}{r_\sigma \tau_\sigma{}'} \\ 0 & 0 \end{bmatrix} \cdot \begin{bmatrix} u_{sd} \\ u_{sq} \end{bmatrix} \tag{7.1}$$

$$\begin{bmatrix} i_{sd} \\ i_{sq} \\ \psi_{rd} \end{bmatrix} = \begin{bmatrix} 1 & 0 & 0 \\ 0 & 1 & 0 \\ 0 & 0 & 1 \end{bmatrix} \cdot \begin{bmatrix} i_{sd} \\ i_{sq} \\ \psi_{rd} \end{bmatrix} + \begin{bmatrix} 0 & 0 \\ 0 & 0 \\ 0 & 0 \end{bmatrix} \cdot \begin{bmatrix} u_{sd} \\ u_{sq} \end{bmatrix} \tag{7.2}$$

Considering a linear, time-invariant system, the values of the elements of the matrices **A**, **B**, **C** and **D** are constant. However, in our case, this is not true for the rotating speeds ω_s and ω, as they change their values during machine operation. Unfortunately, the mathematically correct method, i. e. including ω_s and ω in the input vector $\mathbf{u}$ is not applicable since the rotating speeds are multiplicatively linked with different state variables resulting in a nonlinearity. The state space representation for linear systems cannot represent this fact. As the time constant of the speed control loop has a much higher value than the one of the current control loop, ω_s and ω can be treated as constants for the runtime of the current controller. Hence, ω_s and ω are treated as parameters similar to other machine parameters. The values of the remaining parameters are constant. For the motor used for the experiments, the matrices result (according to appendix D) to

$$\mathbf{A} = \begin{bmatrix} -\frac{1}{\tau_\sigma{}'} & \omega_s & \frac{k_r}{r_\sigma \tau_r \tau_\sigma{}'} \\ -\omega_s & -\frac{1}{\tau_\sigma{}'} & -\frac{k_r \omega}{r_\sigma \tau_\sigma{}'} \\ \frac{l_h}{\tau_r} & 0 & -\frac{1}{\tau_r} \end{bmatrix} = \begin{bmatrix} -0,3964 & \omega_s & 0,07380 \\ -\omega_s & -0,3964 & -4,450\,\omega \\ 0,04245 & 0 & -0,01658 \end{bmatrix} \tag{7.3}$$

$$\mathbf{B} = \begin{bmatrix} \frac{1}{r_\sigma \tau_\sigma{}'} & 0 \\ 0 & \frac{1}{r_\sigma \tau_\sigma{}'} \\ 0 & 0 \end{bmatrix} = \begin{bmatrix} 4,641 & 0 \\ 0 & 4,641 \\ 0 & 0 \end{bmatrix} \tag{7.4}$$

$$\mathbf{C} = \begin{bmatrix} 1 & 0 & 0 \\ 0 & 1 & 0 \\ 0 & 0 & 1 \end{bmatrix} \tag{7.5}$$

$$\mathbf{D} = \begin{bmatrix} 0 & 0 \\ 0 & 0 \\ 0 & 0 \end{bmatrix} \tag{7.6}$$

To use these equations in a model-based control method, they must be transformed into a discrete-time form according to (5.4) and (5.5). For this several procedures exist.

7.1.1 MATLAB

The mathematical computer algebra system *MATLAB*® from *The MathWorks Inc.* is a powerful tool for the calculation of matrices. For an easy treatment of control systems, the additional *Control System Toolbox* is available. With the functions available within this toolbox, complete system models can be transformed from a continuous-time into a discrete-time form and from state space representation into a transfer function-based system model. Unfortunately, MATLAB® cannot handle symbolic mathematics without further extensions, thus, only numerical solutions can be given for the different parameters and elements. Since this is not useable in our case because of the varying parameters ω_s and ω, MATLAB® cannot be used in this way. Hence, the software *Maple*® from *Waterloo Maple Inc.* which is able to handle symbolic mathematics is used for the further work, too.

7.1.2 Difference quotient

The most simple way to transform a continuous-time representation of a system into a discrete-time representation consists of approximating the derivative operator with a difference.

$$\frac{dx}{dt} \approx \frac{x(k+1) - x(k)}{\Delta t}$$

If this is applied to the equations (5.2) and (5.3) and if the constant sampling time T_0 is used for the time difference Δt, the following equations can be obtained:

$$\frac{\mathbf{x}(k+1) - \mathbf{x}(k)}{T_0} = \mathbf{A}\mathbf{x}(k) + \mathbf{B}\mathbf{u}(k) \tag{7.7}$$

$$\mathbf{y}(k) = \mathbf{C}\mathbf{x}(k) + \mathbf{D}\mathbf{u}(k) \tag{7.8}$$

Equation (7.7) can be further transformed into

$$\begin{aligned} \mathbf{x}(k+1) - \mathbf{x}(k) &= T_0\mathbf{A}\mathbf{x}(k) + T_0\mathbf{B}\mathbf{u}(k) \\ \mathbf{x}(k+1) &= (T_0\mathbf{A} + \mathbf{I})\mathbf{x}(k) + T_0\mathbf{B}\mathbf{u}(k) \end{aligned} \tag{7.9}$$

A comparison of (7.9) with (5.4) and of (7.8) with (5.5), respectively, delivers the following expressions for the matrices of the discrete-time representation:

$$\mathbf{A_d} = T_0\mathbf{A} + \mathbf{I} \tag{7.10}$$
$$\mathbf{B_d} = T_0\mathbf{B} \tag{7.11}$$
$$\mathbf{C_d} = \mathbf{C} \tag{7.12}$$
$$\mathbf{D_d} = \mathbf{D} \tag{7.13}$$

If the above transformations are carried out on (7.3)–(7.6), the following results can be obtained:

$$\mathbf{A_d} = \begin{bmatrix} 1-\dfrac{T_0}{\tau_\sigma{}'} & T_0\omega_s & \dfrac{T_0 k_r}{r_\sigma \tau_r \tau_\sigma{}'} \\ -T_0\omega_s & 1-\dfrac{T_0}{\tau_\sigma{}'} & -\dfrac{T_0 k_r \omega}{r_\sigma \tau_\sigma{}'} \\ \dfrac{T_0 l_h}{\tau_r} & 0 & 1-\dfrac{T_0}{\tau_r} \end{bmatrix} = \begin{bmatrix} 0,9872 & 0,03217\ \omega_s & 0,002374 \\ -0,03217\ \omega_s & 0,9872 & -0,1432\ \omega \\ 0,001366 & 0 & 0,9995 \end{bmatrix} \tag{7.14}$$

$$\mathbf{B_d} = \begin{bmatrix} \dfrac{T_0}{r_\sigma \tau_\sigma{}'} & 0 \\ 0 & \dfrac{T_0}{r_\sigma \tau_\sigma{}'} \\ 0 & 0 \end{bmatrix} = \begin{bmatrix} 0,1493 & 0 \\ 0 & 0,1493 \\ 0 & 0 \end{bmatrix} \tag{7.15}$$

$$\mathbf{C_d} = \mathbf{C} \quad \text{(unchanged)} \tag{7.16}$$
$$\mathbf{D_d} = \mathbf{D} \quad \text{(unchanged)} \tag{7.17}$$

7.1.3 Laplace transformation

A mathematically exact calculation of the matrices of the discrete-time state space representation is of course not possible with the simple method described above. For this purpose, the solution of the matrix differential equation (5.2) has to be determined. The following expression is selected analog to the scalar case:

$$\mathbf{x}(t) = e^{\mathbf{A}t}\mathbf{k} \tag{7.18}$$

The matrix $e^{\mathbf{A}t}$ is called *matrix exponential function*. It is defined by the following series [84]:

$$e^{\mathbf{A}t} = \sum_{i=0}^{\infty} \frac{\mathbf{A}^i t^i}{i!} = \mathbf{I} + \mathbf{A}t + \frac{\mathbf{A}^2}{2!}t^2 + \frac{\mathbf{A}^3}{3!}t^3 + \frac{\mathbf{A}^4}{4!}t^4 + \cdots \tag{7.19}$$

From (5.2) and with the expression (7.18) the equation of motion

$$\mathbf{x}(t) = \mathbf{\Phi}(t)\mathbf{x_0} + \int_0^t \mathbf{\Phi}(t-\tau)\mathbf{B}\mathbf{u}(\tau)d\tau \tag{7.20}$$

with

$$\mathbf{\Phi}(t) = e^{\mathbf{A}t} \tag{7.21}$$

can be obtained. $\mathbf{\Phi}$ is called *transition matrix*. If the time is set to $t = (k+1)T_0$, the following equation results from (7.20) with the help of definition (7.21):

$$\mathbf{x}((k+1)T_0) = e^{\mathbf{A}(k+1)T_0}\mathbf{x_0} + \int_0^{(k+1)T_0} e^{\mathbf{A}((k+1)T_0-\tau)}\mathbf{B}\mathbf{u}(\tau)d\tau \tag{7.22}$$

In a digital control the values of the actuating variables are only changed at the sampling instances. Hence, the value of $\mathbf{u}(t)$ is constant for $kT_0 \leqq t \leqq (k+1)T_0$. Thus, the following equation results from (7.22) after some intermediate steps and the substitution $\alpha = (k+1)T_0 - \tau$ [85]:

$$\mathbf{x}((k+1)T_0) = e^{\mathbf{A}T_0}\mathbf{x}(kT_0) + \int_0^{T_0} e^{\mathbf{A}\alpha}d\alpha\ \mathbf{B}\mathbf{u}(kT_0) \tag{7.23}$$

By comparing the coefficients of the above equation with (5.4), $\mathbf{A_d}$ and $\mathbf{B_d}$ can be determined directly:

$$\mathbf{A_d} = e^{\mathbf{A}T_0} \tag{7.24}$$

$$\mathbf{B_d} = \int_0^{T_0} e^{\mathbf{A}\alpha}d\alpha\ \mathbf{B} \tag{7.25}$$

Assuming a continuous system, it can be stated that $\det \mathbf{A} \neq 0$; the input matrix of the discrete-time system thus results to:

$$\mathbf{B_d} = \mathbf{A}^{-1}\left(e^{\mathbf{A}T_0} - \mathbf{I}\right)\mathbf{B} \tag{7.26}$$

With the help of the equations (7.24) and (7.26), the state matrix $\mathbf{A_d}$ and the input matrix $\mathbf{B_d}$ of the discrete-time system can be calculated. Since the output equation (5.3) does not contain derivative operators, it can easily be transformed into the discrete form with the substitution $t = kT_0$. Comparing it to (5.5), it can be seen that $\mathbf{C_d} = \mathbf{C}$ and $\mathbf{D_d} = \mathbf{D}$ does always apply.

For the transformation of the continuous-time state space representation into the discrete-time one, it is necessary to calculate the matrix exponential function $e^{\mathbf{A}t}$. Besides the series expansion given in equation (7.19), Schwarz [108] shows that $e^{\mathbf{A}t}$ can also be determined with the help of the Laplace transformation:

$$e^{\mathbf{A}t} = \mathfrak{L}^{-1}\left\{(s\mathbf{I} - \mathbf{A})^{-1}\right\} \tag{7.27}$$

Since Maple® contains functions for the execution of Laplace operations the above transformation can be implemented easily. However, an analytic calculation shows that in this case the state and input matrix of the discrete-time representation require more than 30 DIN A 4 pages. Even though some truncations and simplifications are certainly possible, the result can surely not be implemented in a real time control due to its extremely high computation time. Thus, the analytically exact calculation of the discrete-time representation by means of Laplace transformation cannot be considered to be feasible anymore.

7.1.4 Power series

The third possibility for the calculation of the matrix exponential function is by truncating the series expansion from equation (7.19) after the desired accuracy is obtained. If the series expansion is truncated after the first element, the following expression can be obtained:

$$e^{\mathbf{A}T_0} \approx T_0\mathbf{A} + \mathbf{I} \tag{7.28}$$

Mathematically, this corresponds to a substitution of the derivative operator with a difference, as shown by a substitution of (7.28) in the equations (7.24) and (7.26) and by a comparison with the transformation equations (7.10) and (7.11), which were derived in chapter 7.1.2.

In the case of a truncation after the second element, the calculation equations for the individual matrix elements do of course become more complex. Then the state and input matrix are:

$$
\mathbf{A_d} = \begin{bmatrix}
\frac{T_0^2(r_\sigma\tau_r^2 - \omega_s^2\tau_\sigma'^2 r_\sigma\tau_r^2 + k_r l_h\tau_\sigma') - 2T_0\tau_\sigma' r_\sigma\tau_r^2 + 2\tau_\sigma'^2 r_\sigma\tau_r^2}{2\tau_\sigma'^2 r_\sigma\tau_r^2} & \frac{T_0\omega_s(\tau_\sigma' - T_0)}{\tau_\sigma'} & \frac{T_0k_r(2\tau_\sigma'\tau_r - T_0(\tau_r + \omega_s\omega\tau_\sigma'\tau_r^2 + \tau_\sigma'))}{2\tau_\sigma'^2 r_\sigma\tau_r^2} \\
\frac{T_0^2(2\omega_s r_\sigma\tau_r - k_r\omega l_h) - 2T_0\omega_s r_\sigma\tau_r\tau_\sigma'}{2r_\sigma\tau_r\tau_\sigma'} & \frac{T_0^2 - 2T_0\tau_\sigma' + \tau_\sigma'^2(2 - T_0^2\omega_s^2)}{2\tau_\sigma'^2} & \frac{T_0k_r(T_0\omega(\tau_r + \tau_\sigma') - 2\omega\tau_\sigma'\tau_r - T_0\omega_s\tau_\sigma')}{2\tau_\sigma'^2 r_\sigma\tau_r} \\
\frac{T_0 l_h(2\tau_\sigma'\tau_r - T_0(\tau_r + \tau_\sigma'))}{2\tau_\sigma'\tau_r^2} & \frac{T_0^2 l_h\omega_s}{2\tau_r} & \frac{T_0^2(k_r l_h + r_\sigma\tau_\sigma') - 2T_0 r_\sigma\tau_r\tau_\sigma' + 2r_\sigma\tau_r^2\tau_\sigma'}{2\tau_\sigma' r_\sigma\tau_r^2}
\end{bmatrix}
= \begin{bmatrix}
0,9873 - 0,0005175\,\omega_s^2 & 0,03176\,\omega_s & 0,002358 - 0,002303\,\omega_s\,\omega \\
-0,03176\,\omega_s - 9,776\cdot 10^{-5}\,\omega & 0,9873 - 0,0005175\,\omega_s^2 & -0,1422\,\omega - 3,819\cdot 10^{-5}\,\omega_s \\
0,001357 & 2,197\cdot 10^{-5}\,\omega_s & 0,9995
\end{bmatrix}
\tag{7.29}
$$

$$
\mathbf{B_d} = \begin{bmatrix}
\frac{(2\tau_\sigma' - T_0)T_0}{2\tau_\sigma'^2 r_\sigma} & \frac{\omega_s T_0^2}{2\tau_\sigma' r_\sigma} \\
-\frac{\omega_s T_0^2}{2\tau_\sigma' r_\sigma} & \frac{(2\tau_\sigma' - T_0)T_0}{2\tau_\sigma'^2 r_\sigma} \\
\frac{T_0^2 l_h}{2r_\sigma\tau_\sigma'\tau_r} & 0
\end{bmatrix}
= \begin{bmatrix}
0,1484 & 0,002402\,\omega_s \\
-0,002402\,\omega_s & 0,1484 \\
1,020\cdot 10^{-4} & 0
\end{bmatrix}
\tag{7.30}
$$

7.2 Experimental results

In order to evaluate the accuracy of the different prediction algorithms they were implemented in a conventional field-oriented control with PI controllers and the predicted results were compared to the values obtained from the actual system. Figure 7.1 shows the response when a step in the reference value for the torque-producing component of the stator current i_{sq} was applied, while figure 7.2 shows the system response for a reference value change in the flux.producing component of the stator current component i_{sd}.

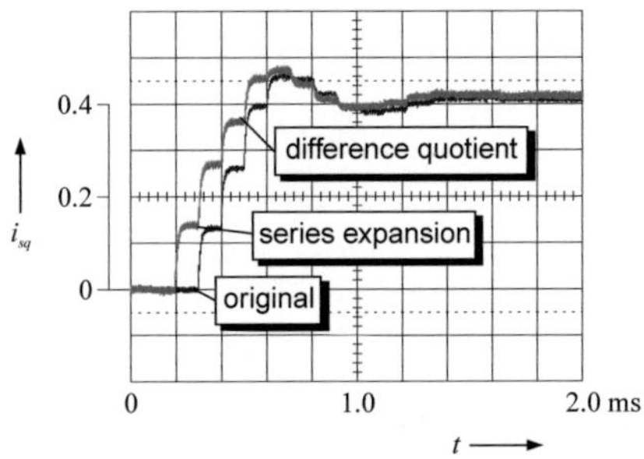

(a) Advance calculation of i_{sq}

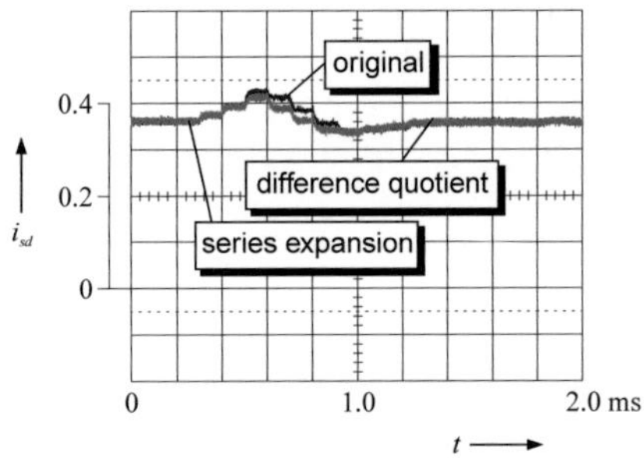

(b) Advance calculation of i_{sd}

Figure 7.1: Prediction with a step change in the reference variable i_{sq}

In figure 7.1(a) the value of the advance calculation for i_{sq} is shown, i. e. the value that will be reached by the current in the *next* sampling cycle. The deviations between the advance calculation and the actual values are relatively small. They are caused by parasitic effects which have not been taken into account when the system model was derived.

The prediction of i_{sd} is shown in figure 7.1(b), the change in the predicted values appears at the *same time* as the actual current rise. At first sight this is astonishing since the prediction should consider the direct influence of u_{sq} on i_{sd}. Nevertheless, equation (7.15) shows that there is *no direct* influence of u_{sq} on i_{sd} when the prediction is done with a simple difference quotient. However, the power series method considers this influence, but with an extremely small value (see equation (7.30)). Obviously a system of higher order is necessary for the exact modeling of the cross coupling; a second order system delivers exactly the same results as the simple difference equation. Nevertheless, the correlation between prediction and reality is in both cases as good as for the prediction of i_{sq} in figure 7.1(a).

Figure 7.2(a) shows the response of i_{sq} when a step in the flux-producing component of the stator current from $i_{sd} = 0.1$ to $i_{sd} = 0.35$ was applied. The torque-producing component of the stator current was controlled to $i_{sq} = 0$. As the step change in i_{sd} does not lead to a measurable influence on i_{sq}, the actual values as well as both predicted values are constantly 0.

As shown in figure 7.2(b), the prediction of the flux-producing component of the stator current i_{sd} when a voltage u_{sd} is applied behaves in the same way as the prediction of the torque-producing component of the stator current i_{sq} when u_{sq} is applied (figure 7.1(a)). The values that will be measured in the *next* sampling cycle are precalculated; hence, the correlation between the values of the actual system and the values calculated with the model is sufficiently good.

All measurements confirm that the calculation of the discrete-time machine model using the series expansion of the matrix exponential function shows no improvement in the prediction compared to the simple approximation via the difference quotient. A comparison of the equations (7.29) and (7.30), resulting from the series expansion with the corresponding equations (7.14) and (7.15) for the difference quotient, shows that the calculation complexity is clearly much higher if series expansion is used. Hence, the method explained in chapter 7.1.2, i. e. modeling via the difference quotient should be preferred.

7.3 Modified machine model for GPC

In practical applications of the machine model derived from the equations (5.4) and (5.5) in chapter 7.1 it can be seen that the approximation of the nonlinearities by treating ω and ω_s as parameters is not feasible. The discrete state

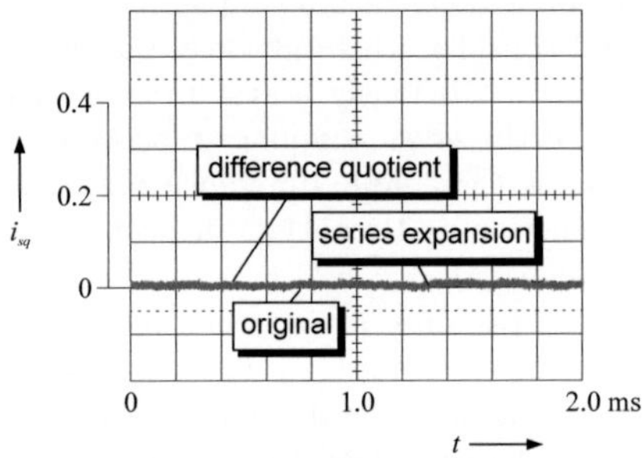

(a) Advance calculation of i_{sq}

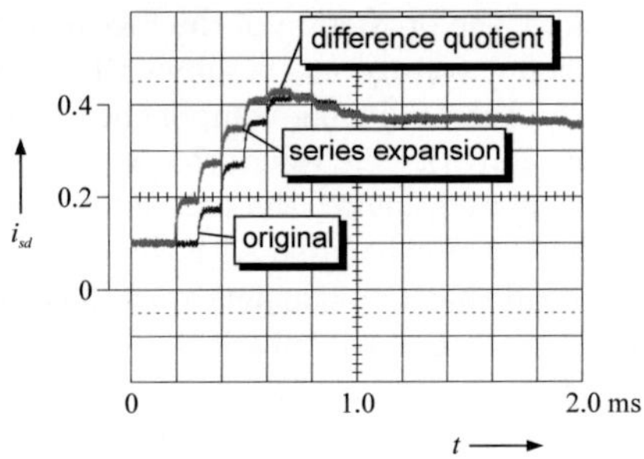

(b) Advance calculation of i_{sd}

Figure 7.2: Prediction with a step change in the reference variable i_{sd}

matrix $\mathbf{A_d}$ would change in every sampling cycle and therefore the various polynomial matrices of the GPC controller, as shown in chapter 8, would also have to be completely calculated again in every sampling cycle. Since this contains, among other things, a matrix inversion, the necessary arithmetic operations exceed the available computation time. Hence, this method is not feasible with the available resources.

An alternative would be to extract the feedback coupling branches from the model and to substitute them with additional system inputs. In this approach from the known values of ω, ω_s, i_{sd}, i_{sq} and ψ_{rd}, three virtual "inputs" $\omega_s i_{sd}$, $\omega_s i_{sq}$ and $\omega\psi_{rd}$ would be calculated, which would then be fed back to the system. In simulations, this works well, in practice, a passively running model also shows good results; however, if this structure should be used for control via GPC, the GPC controller then interprets the additional pseudo inputs $\omega_s i_{sd}$, $\omega_s i_{sq}$ and $\omega\psi_{rd}$, which were added in order to consider the cross coupling and the back EMF, as real plant inputs and consequently the GPC controller delivers values for these actuating variables. However, since these are no real actuating variables, they cannot be influenced by the controller. Thus, a model of this kind *cannot be used* for GPC control.

A solution is to treat the feedback couplings as known disturbance variables and to consider them already in the modeling. Therefore the state space model (5.2) and (5.3)—as described by Camacho/Bordons—is expanded for the consideration of the disturbance inputs [20]

$$\frac{d}{dt}\mathbf{x} = \mathbf{A}\mathbf{x} + \mathbf{B}\mathbf{u} + \mathbf{E}\mathbf{v} \tag{7.31}$$

$$\mathbf{y} = \mathbf{C}\mathbf{x} + \mathbf{D}\mathbf{u} + \mathbf{F}\mathbf{v} \tag{7.32}$$

The *input matrix for disturbance variables* $\mathbf{E}$ and the *feedforward matrix for disturbance variables* $\mathbf{F}$ are added. The vectors $\mathbf{x}$, $\mathbf{u}$, $\mathbf{v}$ and $\mathbf{y}$ are defined according to the machine model (2.11)–(2.13), in which in contrast to chapter 7.1 the output vector $\mathbf{y}$ does *not* contain the flux ψ_{rd}. Otherwise the GPC controller would treat ψ_{rd} as a real output of the plant and expect a reference value for this variable. Since the flux should not be controlled and only a constant flux-producing current i_{sd} will be given, this is redundant. The definitions of the individual vectors are as follows:

$$\mathbf{x} = \begin{bmatrix} i_{sd} \\ i_{sq} \\ \psi_{rd} \end{bmatrix}, \qquad \mathbf{u} = \begin{bmatrix} u_{sd} \\ u_{sq} \end{bmatrix}, \qquad \mathbf{v} = \begin{bmatrix} \omega_s i_{sd} \\ \omega_s i_{sq} \\ \omega\psi_{rd} \end{bmatrix}, \qquad \mathbf{y} = \begin{bmatrix} i_{sd} \\ i_{sq} \end{bmatrix}$$

whereby the state space representation is given by

$$
\begin{aligned}
\frac{d}{d\tau}\begin{bmatrix} i_{sd} \\ i_{sq} \\ \psi_{rd} \end{bmatrix} &= \begin{bmatrix} -\frac{1}{\tau_\sigma{}'} & 0 & \frac{k_r}{r_\sigma \tau_r \tau_\sigma{}'} \\ 0 & -\frac{1}{\tau_\sigma{}'} & 0 \\ \frac{l_h}{\tau_r} & 0 & -\frac{1}{\tau_r} \end{bmatrix} \cdot \begin{bmatrix} i_{sd} \\ i_{sq} \\ \psi_{rd} \end{bmatrix} \\
&+ \begin{bmatrix} \frac{1}{r_\sigma \tau_\sigma{}'} & 0 \\ 0 & \frac{1}{r_\sigma \tau_\sigma{}'} \\ 0 & 0 \end{bmatrix} \cdot \begin{bmatrix} u_{sd} \\ u_{sq} \end{bmatrix} \\
&+ \begin{bmatrix} 0 & 1 & 0 \\ -1 & 0 & -\frac{k_r}{r_\sigma \tau_\sigma{}'} \\ 0 & 0 & 0 \end{bmatrix} \cdot \begin{bmatrix} \omega_s i_{sd} \\ \omega_s i_{sq} \\ \omega \psi_{rd} \end{bmatrix}
\end{aligned}
\tag{7.33}
$$

$$
\begin{aligned}
\begin{bmatrix} i_{sd} \\ i_{sq} \end{bmatrix} &= \begin{bmatrix} 1 & 0 & 0 \\ 0 & 1 & 0 \end{bmatrix} \cdot \begin{bmatrix} i_{sd} \\ i_{sq} \\ \psi_{rd} \end{bmatrix} \\
&+ \begin{bmatrix} 0 & 0 \\ 0 & 0 \end{bmatrix} \cdot \begin{bmatrix} u_{sd} \\ u_{sq} \end{bmatrix} \\
&+ \begin{bmatrix} 0 & 0 & 0 \\ 0 & 0 & 0 \end{bmatrix} \cdot \begin{bmatrix} \omega_s i_{sd} \\ \omega_s i_{sq} \\ \omega \psi_{rd} \end{bmatrix}
\end{aligned}
\tag{7.34}
$$

So the matrices **A**, **B**, **C** and **D** as well as **E** and **F** result to:

$$
\mathbf{A} = \begin{bmatrix} -\frac{1}{\tau_\sigma{}'} & 0 & \frac{k_r}{r_\sigma \tau_r \tau_\sigma{}'} \\ 0 & -\frac{1}{\tau_\sigma{}'} & 0 \\ \frac{l_h}{\tau_r} & 0 & -\frac{1}{\tau_r} \end{bmatrix} = \begin{bmatrix} -0.3964 & 0 & 0.07380 \\ 0 & -0.3964 & 0 \\ 0.04245 & 0 & -0.01658 \end{bmatrix} \tag{7.35}
$$

$$
\mathbf{B} = \begin{bmatrix} \frac{1}{r_\sigma \tau_\sigma{}'} & 0 \\ 0 & \frac{1}{r_\sigma \tau_\sigma{}'} \\ 0 & 0 \end{bmatrix} = \begin{bmatrix} 4.641 & 0 \\ 0 & 4.641 \\ 0 & 0 \end{bmatrix} \tag{7.36}
$$

$$\mathbf{C} = \begin{bmatrix} 1 & 0 & 0 \\ 0 & 1 & 0 \end{bmatrix} \tag{7.37}$$

$$\mathbf{D} = \begin{bmatrix} 0 & 0 \\ 0 & 0 \end{bmatrix} \tag{7.38}$$

$$\mathbf{E} = \begin{bmatrix} 0 & 1 & 0 \\ -1 & 0 & -\frac{k_r}{r_\sigma \tau_\sigma{}'} \\ 0 & 0 & 0 \end{bmatrix} = \begin{bmatrix} 0 & 1 & 0 \\ -1 & 0 & -4.450 \\ 0 & 0 & 0 \end{bmatrix} \tag{7.39}$$

$$\mathbf{F} = \begin{bmatrix} 0 & 0 & 0 \\ 0 & 0 & 0 \end{bmatrix} \tag{7.40}$$

Since a discrete-time model is necessary for the control of the plant with an MPC controller, the equations derived above have to be transferred into a discrete-time state space representation:

$$\mathbf{x}(k+1) = \mathbf{A_d}\mathbf{x}(k) + \mathbf{B_d}\mathbf{u}(k) + \mathbf{E_d}\mathbf{v}(k) \tag{7.41}$$

$$\mathbf{y}(k) = \mathbf{C_d}\mathbf{x}(k) + \mathbf{D_d}\mathbf{u}(k) + \mathbf{F_d}\mathbf{v}(k) \tag{7.42}$$

The matrices $\mathbf{E_d}$ and $\mathbf{F_d}$ thereby represent the input and feedforward matrix for the discrete-time disturbance function $\mathbf{v}(k)$.

As all system matrices now contain no symbolic parameters, but only numerical values, an analytic calculation of $\mathbf{A_d}$, $\mathbf{B_d}$, $\mathbf{C_d}$, $\mathbf{D_d}$, $\mathbf{E_d}$ and $\mathbf{F_d}$ is not required anymore. Therefore, a method programmed with the help of Maple® in order to determine the matrices of the discrete-time state space representation via difference quotient (chapter 7.1.2), Laplace transformation (chapter 7.1.3) or power series expansion (chapter 7.1.4) is not necessary. The calculation of the matrices of the equations (7.41) and (7.42) can easily be done with MATLAB®, resulting in the following values:

$$\mathbf{A_d} = \begin{bmatrix} 0.9873 & 0 & 0.002358 \\ 0 & 0.9873 & 0 \\ 0.001357 & 0 & 0.9995 \end{bmatrix}$$

$$\mathbf{B_d} = \begin{bmatrix} 0.1484 & 0 \\ 0 & 0.1484 \\ 0.0001015 & 0 \end{bmatrix}$$

$$\mathbf{C_d} = \mathbf{C} \qquad \text{unchanged}$$

$$\mathbf{D_d} = \mathbf{D} \qquad \text{unchanged}$$

$$\mathbf{E_d} = \begin{bmatrix} 0 & 0.03197 & 0 \\ -0.03197 & 0 & -0.1423 \\ 0 & 2.187 \cdot 10^{-5} & 0 \end{bmatrix}$$

$$\mathbf{F_d} = \mathbf{F} \qquad \text{unchanged}$$

It is remarkable that $\mathbf{E_d}_{3,2} = 2.187 \cdot 10^{-5}$, even though it is expected to be equal to zero: $\mathbf{E_d}_{3,2} = 0$. Obviously internal rounding errors of MATLAB® lead to this deviation. The difference is relatively small, however, it can of course be corrected when the model is integrated into a control structure.

8 Multivariable GPC control

As MPC controllers are in principle already designed as multidimensional controllers, the extension of a SISO-GPC to a multivariable method does not involve too much complexity and costs. Hence, the derivation nearly exactly corresponds to the one presented in chapter 6.1.1 for a SISO system, however, due to the higher dimensions, vectors become matrices and polynomials become polynomial matrices. Besides, it has to be noted that special mathematical rules apply for polynomial matrices and matrix polynomials. Good summaries of these rules can be found in the publication by Goodwin/Sin [44] and on the website of the company PolyX [125]. Appendix A gives an overview about some of the most important mathematical rules. Further information about matrix theory can be taken from the books by Gantmacher [39], Lancaster/Tismenetsky [75] and Zurmühl/Falk [124].

8.1 "Classical" MIMO-GPC

The "classical" multidimensional GPC is a simple extension of the uni-dimensional method to a multidimensional one.

8.1.1 Determination of the transfer function

Before an MPC based on a transfer function-based model can be designed, the state space representation (equations (5.4) and (5.5)) must be converted into a transfer function and/or matrix. If the transformation is done for a unidimensional control, a discrete-time transfer function of the form

$$G_d(z^{-1}) = g_{d0} + g_{d1}z^{-1} + g_{d2}z^{-2} + \cdots + g_{dn}z^{-n} = \frac{Y(z)}{U(z)}$$

results. However, for a multidimensional control the division $\mathbf{Y}(z)/\mathbf{U}(z)$ results into a matrix polynomial, i. e. a polynomial whose coefficients are real matrices.

$$\mathbf{G_d}(z^{-1}) = \mathbf{G_{d0}} + \mathbf{G_{d1}}z^{-1} + \mathbf{G_{d2}}z^{-2} + \cdots + \mathbf{G_{dn}}z^{-n} = \frac{\mathbf{Y}(z)}{\mathbf{U}(z)} \tag{8.1}$$

For the individual elements of $\mathbf{Y}(z)$ and $\mathbf{U}(z)$, the following applies:

$$\begin{aligned}&\mathbf{y}(t) : \text{Vector}(n \times 1) &&\text{with } n = \text{number of system outputs}\\ &\mathbf{u}(t) : \text{Vector}(m \times 1) &&\text{with } m = \text{number of system inputs}\end{aligned}$$

Because of this, for the discrete-time transfer function $\mathbf{G_d}(z^{-1})$, the following applies, too:

$$\mathbf{G_d}(z^{-1}) : \text{Matrix}(n \times m) \qquad (\text{rows} \times \text{columns})$$

Matrix polynomials can be transferred into polynomial matrices and vice versa [39]; polynomial matrices are matrices whose elements are polynomials. Considering the polynomial matrix $\mathbf{P}(z^{-1})$:

$$\mathbf{P}(z^{-1}) = \left\|P_{j,i}(z^{-1})\right\|^{n,m} = \left\|p_{j,i_0} + p_{j,i_1}z^{-1} + \cdots + p_{j,i_k}z^{-k}\right\|^{n,m}$$

The polynomial matrix $\mathbf{P}(z^{-1})$ can be written as a polynomial in z with the coefficients being real matrices:

$$\mathbf{P}(z^{-1}) = \mathbf{P}_0 + \mathbf{P}_1 z^{-1} + \cdots + \mathbf{P}_k z^{-k}$$

with

$$\mathbf{P}_x = \left\|p_{j,i_x}\right\|^{n,m}$$

Therefore, a polynomial matrix $\mathbf{P}(z^{-1})$ can be transferred into a matrix polynomial and vice versa.

For the calculation of the transfer matrix $\mathbf{G_d}(z^{-1})$, the state space representation (equations (5.4) and (5.5)) is considered first:

$$\begin{aligned}\mathbf{x}(k+1) &= \mathbf{A_d}\mathbf{x}(k) + \mathbf{B_d}\mathbf{u}(k)\\ \mathbf{y}(k) &= \mathbf{C_d}\mathbf{x}(k) + \mathbf{D_d}\mathbf{u}(k)\end{aligned}$$

These equations are then transformed into the Z-domain:

$$z\mathbf{X}(z) = \mathbf{A_d}\mathbf{X}(z) + \mathbf{B_d}\mathbf{U}(z) \tag{8.2}$$

$$\mathbf{Y}(z) = \mathbf{C_d}\mathbf{X}(z) + \mathbf{D_d}\mathbf{U}(z) \tag{8.3}$$

By rearranging (8.2), the following equation can be obtained:

$$\mathbf{X}(z) = (z\mathbf{I} - \mathbf{A_d})^{-1}\mathbf{B_d}\mathbf{U}(z)$$

Replacing it into (8.3) results in

$$\mathbf{Y}(z) = \big(\mathbf{C_d}(z\mathbf{I} - \mathbf{A_d})^{-1}\mathbf{B_d} + \mathbf{D_d}\big)\mathbf{U}(z)$$

Thus, the transfer function $\mathbf{G_d}(z^{-1})$ can be written as

$$\mathbf{G_d}(z^{-1}) = \mathbf{C_d}(z\mathbf{I} - \mathbf{A_d})^{-1}\mathbf{B_d} + \mathbf{D_d} \tag{8.4}$$

8.1.2 Calculation of the system matrices

In the case of a unidimensional control, the system polynomials $A(z^{-1})$ and $B(z^{-1})$, on which the CARIMA model is based, can easily be taken from the transfer function $G_d(z^{-1})$. As shown in equation (6.1), $A(z^{-1})$ is the denominator and $B(z^{-1})$ is the numerator of $G_d(z^{-1})$.

However, if it is a multidimensional system, then $\mathbf{A}(z^{-1})$ and $\mathbf{B}(z^{-1})$ are system *polynomial matrices*. Since $\mathbf{G_d}(z^{-1})$ is also a polynomial matrix, $\mathbf{A}(z^{-1})$ and $\mathbf{B}(z^{-1})$ cannot simply be determined by taking the denominator and numerator of $\mathbf{G_d}(z^{-1})$. A simple method for the determination of $\mathbf{A}(z^{-1})$ and $\mathbf{B}(z^{-1})$ is presented by Camacho in [20, chapter 6.2.1]. Furthermore a mathematically more complex method which is also described by Goodwin and Sin [44, chapter 2.3.5] is explained which delivers left coprime[1] results for $\mathbf{A}(z^{-1})$ and $\mathbf{B}(z^{-1})$. Although this is not necessary for the implementation of a multivariable GPC controller, using left coprime system matrices results in more efficient algorithms leading to shorter computation times. This procedure is explained in detail in appendix G; a very good and detailed description can also be found in the publication by Geering [41], this is the reason why further explanations are omitted at this point. For the sake of simplicity, only the simpler method by Camacho [20] is treated in the following.

It is well-known that

$$\mathbf{G_d}(z^{-1}) = \mathbf{A}(z^{-1})^{-1}\mathbf{B}(z^{-1})z^{-1} \tag{8.5}$$

Now the most simple way is to assume that $\mathbf{A}(z^{-1})$ is a diagonal matrix whose diagonal elements are the least common multiples of the denominators of the corresponding rows of $\mathbf{G_d}(z^{-1})$. Then, the polynomial matrix $\mathbf{B}(z^{-1})$ can simply be calculated with

$$\mathbf{B}(z^{-1}) = \mathbf{A}(z^{-1})\mathbf{G_d}(z^{-1})z \tag{8.6}$$

[1] See appendix A for an explanation of the mathematical terms for polynomial matrices.

8.1.3 Mathematical derivation

The CARIMA model

Analog to SISO systems, the following approach is taken for the CARIMA model:

$$\mathbf{A}(z^{-1})\mathbf{y}(t) = \mathbf{B}(z^{-1})\mathbf{u}(t-1) + \mathbf{C}(z^{-1})\frac{\boldsymbol{\xi}(t)}{\Delta} \tag{8.7}$$

in which now the following applies:

$$\begin{aligned}
\mathbf{A}(z^{-1}) &= \mathbf{I} + \mathbf{A}_1 z^{-1} + \mathbf{A}_2 z^{-2} + \cdots + \mathbf{A}_{na} z^{-na} \\
\mathbf{B}(z^{-1}) &= \mathbf{B}_0 + \mathbf{B}_1 z^{-1} + \mathbf{B}_2 z^{-2} + \cdots + \mathbf{B}_{nb} z^{-nb} \\
\mathbf{C}(z^{-1}) &= \mathbf{I} + \mathbf{C}_1 z^{-1} + \mathbf{C}_2 z^{-2} + \cdots + \mathbf{C}_{nc} z^{-nc} \\
\Delta &= 1 - z^{-1}
\end{aligned}$$

The dimensions of the individual matrices are:

$$\begin{aligned}
\mathbf{A}(z^{-1}) &: \text{Matrix}(n \times n) \\
\mathbf{B}(z^{-1}) &: \text{Matrix}(n \times m) \\
\mathbf{C}(z^{-1}) &: \text{Matrix}(n \times n) \\
\boldsymbol{\xi}(t) &: \text{Vector}(n \times 1)
\end{aligned}$$

In this case, the matrix polynomial can also be set to $\mathbf{C}(z^{-1}) = \mathbf{I}$ if $\boldsymbol{\xi}(t)$ represents white noise. Then equation (8.7) simplifies to

$$\mathbf{A}(z^{-1})\mathbf{y}(t) = \mathbf{B}(z^{-1})\mathbf{u}(t-1) + \frac{\boldsymbol{\xi}(t)}{\Delta} \tag{8.8}$$

The j-step ahead predictor

As in 6.1.1, a Diophantine equation is used for the derivation of an optimum predictor, too:

$$\begin{aligned}
\mathbf{I} &= \mathbf{E}_j(z^{-1})\mathbf{A}(z^{-1})\Delta + z^{-j}\mathbf{F}_j(z^{-1}) \\
&= \mathbf{E}_j(z^{-1})\tilde{\mathbf{A}}(z^{-1}) + z^{-j}\mathbf{F}_j(z^{-1})
\end{aligned} \tag{8.9}$$

with

$$\begin{aligned}
\tilde{\mathbf{A}}(z^{-1}) &= \Delta\mathbf{A}(z^{-1}) \\
\mathbf{E}_j(z^{-1}) &= \mathbf{E}_{j,0} + \mathbf{E}_{j,1} z^{-1} + \mathbf{E}_{j,2} z^{-2} + \cdots + \mathbf{E}_{j,j-1} z^{-(j-1)} \\
\mathbf{F}_j(z^{-1}) &= \mathbf{F}_{j,0} + \mathbf{F}_{j,1} z^{-1} + \mathbf{F}_{j,2} z^{-2} + \cdots + \mathbf{F}_{j,na} z^{-na}
\end{aligned}$$

with the dimensions

$$\mathbf{E}_j(z^{-1}) : \text{Matrix}(n \times n)$$
$$\mathbf{F}_j(z^{-1}) : \text{Matrix}(n \times n)$$

Concerning the degree of the matrix polynomials, the statements given for unidimensional systems apply analogously.

After solving (8.9) for $\mathbf{E}_j(z^{-1})\tilde{\mathbf{A}}(z^{-1})$ and after inserting this result into equation (8.8), which was multiplied with $\Delta\mathbf{E}_j(z^{-1})z^j$ the prediction equation

$$\mathbf{y}(t+j) = \mathbf{E}_j(z^{-1})\mathbf{B}(z^{-1})\Delta\mathbf{u}(t+j-1) + \mathbf{F}_j(z^{-1})\mathbf{y}(t) + \mathbf{E}_j(z^{-1})\boldsymbol{\xi}(t+j)$$

can be obtained. All noise terms are in the future, too, which means that they are unknown and so they can be neglected for the prediction. Hence, the best possible prediction is

$$\hat{\mathbf{y}}(t+j) = \mathbf{E}_j(z^{-1})\mathbf{B}(z^{-1})\Delta\mathbf{u}(t+j-1) + \mathbf{F}_j(z^{-1})\mathbf{y}(t) \tag{8.10}$$

Different from the SISO-GPC case,

$$\mathbf{E}_j(z^{-1})\mathbf{B}(z^{-1}) = \mathbf{G}_j(z^{-1}) + z^{-j}\mathbf{G}_{jp}(z^{-1}) \qquad \text{degree}\big(\mathbf{G}_j(z^{-1})\big) < j \tag{8.11}$$

is set, whereas the $\mathbf{G}_j(z^{-1})$ matrix polynomials, in contrast to the $G_j(z^{-1})$ polynomials in the unidimensional case only refer to the future. The part forming the free response is completely inside the $z^{-j}\mathbf{G}_{jp}(z^{-1})$ term. Thus, the prediction equation becomes to

$$\begin{aligned}\hat{\mathbf{y}}(t+j) = {} & \underbrace{\mathbf{G}_j(z^{-1})\Delta\mathbf{u}(t+j-1)}_{\text{forced response}} \\ & + \underbrace{\mathbf{G}_{jp}(z^{-1})\Delta\mathbf{u}(t-1) + \mathbf{F}_j(z^{-1})\mathbf{y}(t)}_{\substack{\text{all parameters are in the past} \\ \rightarrow \text{ free response } \mathbf{f}(t+j)}}\end{aligned} \tag{8.12}$$

$$\hat{\mathbf{y}}(t+j) = \mathbf{G}_j(z^{-1})\Delta\mathbf{u}(t+j-1) + \mathbf{f}(t+j) \tag{8.13}$$

with

$$\mathbf{f}(t+j) = \mathbf{G}_{jp}(z^{-1})\Delta\mathbf{u}(t-1) + \mathbf{F}_j(z^{-1})\mathbf{y}(t)$$

with the dimensions

$$\mathbf{G}_j(z^{-1}), \mathbf{G}_{jp}(z^{-1}) : \text{Matrix}(n \times m)$$
$$\mathbf{f}(t) : \text{Vector}(n \times 1)$$

Therefore, the single prediction steps for a MIMO system are:

$$\begin{aligned}
\hat{\mathbf{y}}(t+1) &= \mathbf{G}_1(z^{-1})\Delta\mathbf{u}(t) + \mathbf{f}(t+1) \\
\hat{\mathbf{y}}(t+2) &= \mathbf{G}_2(z^{-1})\Delta\mathbf{u}(t+1) + \mathbf{f}(t+2) \\
\hat{\mathbf{y}}(t+3) &= \mathbf{G}_3(z^{-1})\Delta\mathbf{u}(t+2) + \mathbf{f}(t+3) \\
&\vdots \\
\hat{\mathbf{y}}(t+N_p) &= \mathbf{G}_{N_p}(z^{-1})\Delta\mathbf{u}(t+N_p-1) + \mathbf{f}(t+N_p)
\end{aligned}$$

Calculation of the actuating variables

A quadratic cost function is applied to the values precalculated with (8.13) in order to calculate an optimum sequence of values for the actuating variables. The cost function is the same as the one for unidimensional control.

$$J(N_1, N_2, N_u) = \sum_{j=N_1}^{N_2} \mu_j \left\|\hat{\mathbf{y}}(t+j) - \mathbf{w}(t+j)\right\|^2 + \sum_{j=1}^{N_u} \lambda_j \left\|\Delta\mathbf{u}(t+j-1)\right\|^2 \quad (8.14)$$

Because of calculating the absolute value and because of squaring the expected control error and, respectively, future changes of the actuating variables, only a scalar value remains. Hence, the weighting factors μ_j and λ_j are even in the multidimensional case scalar values, too. Unfortunately, a different weighting of the different system outputs is, because of this, not possible because then the control errors or the $\Delta\mathbf{u}$-values would have to be multiplied with a weighing matrix before the magnitude is calculated.

According to the unidimensional case, $\mu_j = 1$ for all j is selected and mostly also $N_1 = 1$ and $N_2 = N_p$ is set.

To optimize the values of the actuating variables, the term $\mathbf{G}_{jp}(z^{-1})\Delta\mathbf{u}(t-1)$ in equation (8.13) is combined to a multiplication of real matrices $\mathbf{G}\tilde{\mathbf{U}}$. Thus, similar to the SISO system, the following can be written:

$$\mathbf{Y} = \mathbf{G}\tilde{\mathbf{U}} + \mathbf{F} \quad (8.15)$$

in which also $\tilde{\mathbf{U}} = \Delta\mathbf{U}$. Analog to the unidimensional case, a simplified MIMO cost function corresponding to the unidimensional equation (6.8) can be set up by substituting (8.15) in (8.14):

$$J = \tilde{\mathbf{U}}^T(\mathbf{G}^T\mathbf{G} + \lambda\mathbf{I})\tilde{\mathbf{U}} + 2\tilde{\mathbf{U}}^T\mathbf{G}^T(\mathbf{F} - \mathbf{W}) + (\mathbf{F} - \mathbf{W})^T(\mathbf{F} - \mathbf{W}) \qquad (8.16)$$

After the derivation of (8.16) with respect to $\frac{\partial}{\partial\tilde{\mathbf{U}}}$ and by setting the resulting term equal to zero, optimum future values for the actuating variables according to

$$\tilde{\mathbf{U}} = (\mathbf{G}^T\mathbf{G} + \lambda\mathbf{I})^{-1} \cdot \mathbf{G}^T(\mathbf{W} - \mathbf{F}) \qquad (8.17)$$

can be determined, in which the simplifications $\mu_j = 1$ and $\lambda_j = \lambda$ for all j were also applied.

Recursion of the Diophantine equation

The Diophantine equation (8.9) is also solved similar to the SISO system by setting it up for j and $j+1$ and then by subtracting the resulting expressions from each other:

$$\begin{array}{r|l} - & \mathbf{I} = \mathbf{E}_j(z^{-1})\tilde{\mathbf{A}}(z^{-1}) + z^{-j}\mathbf{F}_j(z^{-1}) \\ + & \mathbf{I} = \mathbf{E}_{j+1}(z^{-1})\tilde{\mathbf{A}}(z^{-1}) + z^{-(j+1)}\mathbf{F}_{j+1}(z^{-1}) \\ \hline & 0 = \big(\mathbf{E}_{j+1}(z^{-1}) - \mathbf{E}_j(z^{-1})\big)\tilde{\mathbf{A}}(z^{-1}) + z^{-j}\big(z^{-1}\mathbf{F}_{j+1}(z^{-1}) - \mathbf{F}_j(z^{-1})\big) \end{array}$$

Analog to equation (6.11), the approach

$$\mathbf{E}_{j+1}(z^{-1}) - \mathbf{E}_j(z^{-1}) = \tilde{\mathbf{R}}(z^{-1}) + \mathbf{R}_j z^{-j} \qquad (8.18)$$

is taken because of the degrees of the matrix polynomials. In (8.18) $\tilde{\mathbf{R}}(z^{-1})$ is a matrix polynomial of the degree $\leqq j-1$ and $\mathbf{R}_j$ is a real matrix. As in the unidimensional case, here it also applies that $\tilde{\mathbf{R}}(z^{-1}) = 0$, because $\tilde{\mathbf{A}}(z^{-1})$ is monic[2] ($\tilde{\mathbf{A}}_0 = \mathbf{I}$). After substituting this into the Diophantine equation the following results:

$$z^{-1}\mathbf{F}_{j+1}(z^{-1}) = \mathbf{F}_j(z^{-1}) - \mathbf{R}_j\tilde{\mathbf{A}}(z^{-1})$$

Since, as already mentioned, $\tilde{\mathbf{A}}(z^{-1})$ is monic, the following results after a coefficient comparison of the above matrix polynomial equation:

$$\mathbf{F}_{j,0} = \mathbf{R}_j \qquad (8.19)$$

$$\mathbf{F}_{j+1,i} = \mathbf{F}_{j,i+1} - \mathbf{R}_j\tilde{\mathbf{A}}_{i+1} \qquad (8.20)$$

[2] For an explanation see appendix A.

in which i varies from $0 \ldots \text{degree}(\mathbf{F}_{j+1})$. Moreover, from the equations (8.18), (8.19) and the identity $\tilde{\mathbf{R}}(z^{-1}) = 0$ the recursion law for $\mathbf{E}_j(z^{-1})$ follows:

$$\mathbf{E}_{j+1}(z^{-1}) = \mathbf{E}_j(z^{-1}) + \mathbf{F}_{j,0} z^{-j} \tag{8.21}$$

Considering that $\mathbf{A}(z^{-1})$ is monic the final terms of the recursion are obtained from (8.9) for $j = 1$:

$$\mathbf{E}_1(z^{-1}) = \mathbf{E}_0 = \mathbf{I} \tag{8.22}$$

$$\mathbf{F}_1(z^{-1}) = z\big(\mathbf{I} - \tilde{\mathbf{A}}(z^{-1})\big) \tag{8.23}$$

In the multidimensional case, as it can be seen easily from (8.21), the coefficients from 0 to $-(j-1)$ of the matrix polynomial $\mathbf{E}_{j+1}(z^{-1})$ are all identical, too. Therefore, the individual polynomials can be written as:

$$\begin{aligned}
\mathbf{E}_1(z^{-1}) &= \mathbf{E}_0 \\
\mathbf{E}_2(z^{-1}) &= \mathbf{E}_0 + \mathbf{E}_1 z^{-1} \\
\mathbf{E}_3(z^{-1}) &= \mathbf{E}_0 + \mathbf{E}_1 z^{-1} + \mathbf{E}_2 z^{-2} \\
&\vdots \\
\mathbf{E}_j(z^{-1}) &= \mathbf{E}_0 + \mathbf{E}_1 z^{-1} + \mathbf{E}_2 z^{-2} + \cdots + \mathbf{E}_{j-1} z^{-(j-1)} \\
\mathbf{E}_{j+1}(z^{-1}) &= \mathbf{E}_0 + \mathbf{E}_1 z^{-1} + \mathbf{E}_2 z^{-2} + \cdots + \mathbf{E}_{j-1} z^{-(j-1)} + \mathbf{E}_j z^{-j} \\
&= \mathbf{E}_j(z^{-1}) + \mathbf{E}_j z^{-j}
\end{aligned}$$

in which

$$\mathbf{E}_j = \mathbf{F}_{j,0}$$

Calculation of the free and forced response

As shown on page 87, the prediction equation (8.10) is separated into the forced and the free response (equations (8.13) and (8.15)). From (8.12) follows that only the part $\mathbf{G}_j(z^{-1})$ is essential for the forced response. Therefore, for the calculation of $\mathbf{G}_j(z^{-1})$ according to (8.11), the approach

$$\mathbf{G}_j(z^{-1}) = \mathbf{E}_j(z^{-1})\mathbf{B}(z^{-1})$$

is taken in which, according to definitions, only the terms of the product $\mathbf{E}_j(z^{-1})\mathbf{B}(z^{-1})$ with degree $\leqq (j-1)$ shall be used. Besides, it has to be noted that for all $\mathbf{E}_{k,j}$, it applies that $\mathbf{E}_{k,j} = \mathbf{E}_j$. Therefore, the polynomial matrix $\mathbf{G}_j(z^{-1})$ can be obtained from the terms of degree $0 \ldots (j-1)$ of the following equation

$$\mathbf{G}_j(z^{-1}) = \left(\mathbf{E}_0 + \mathbf{E}_1 z^{-1} + \cdots + \mathbf{E}_{j-1} z^{-(j-1)}\right)\left(\mathbf{B}_0 + \mathbf{B}_1 z^{-1} + \cdots + \mathbf{B}_{nb} z^{-nb}\right)$$

For $\mathbf{G}_{j+1}(z^{-1})$ the terms $0 \ldots j$ are calculated analogously:

$$\begin{aligned}
\mathbf{G}_{j+1}(z^{-1}) &= \mathbf{E}_{j+1}(z^{-1})\mathbf{B}(z^{-1}) \\
&= \left(\mathbf{E}_j(z^{-1}) + \mathbf{F}_{j,0} z^{-j}\right)\mathbf{B}(z^{-1}) \\
&= \left(\mathbf{E}_0 + \mathbf{E}_1 z^{-1} + \cdots + \mathbf{E}_{j-1} z^{-(j-1)}\right)\mathbf{B}(z^{-1}) + \mathbf{F}_{j,0} z^{-j}\mathbf{B}(z^{-1}) \\
&= \mathbf{G}_j(z^{-1}) \\
&\quad + \underbrace{\mathbf{E}_0\mathbf{B}_j z^{-j} + \mathbf{E}_1\mathbf{B}_{j-1} z^{-j} + \cdots + \mathbf{E}_{j-1}\mathbf{B}_1 z^{-j}}_{\substack{\text{additionally from } \left(\mathbf{E}_0+\mathbf{E}_1 z^{-1}+\cdots+\mathbf{E}_{j-1} z^{-(j-1)}\right)\mathbf{B}(z^{-1}) \\ \text{since for } \mathbf{G}_{j+1}(z^{-1}) \text{ only terms up to degree } j \\ \text{need to be considered in the calculation}}} + \mathbf{F}_{j,0}\mathbf{B}_0 z^{-j} \\
&= \mathbf{G}_j(z^{-1}) + \left(\mathbf{F}_{j,0}\mathbf{B}_0 + \mathbf{E}_{j-1}\mathbf{B}_1 + \cdots + \mathbf{E}_0\mathbf{B}_j\right) z^{-j}
\end{aligned}$$

As in unidimensional GPC, the individual $\mathbf{G}_j$ terms differ only in the newly added element, too. Hence it can be written

$$\mathbf{G}_j(z^{-1}) = \mathbf{G}_0 + \mathbf{G}_1 z^{-1} + \mathbf{G}_2 z^{-2} + \cdots + \mathbf{G}_{j-1} z^{-(j-1)} \qquad j = 1 \ldots N_p$$

with

$$\mathbf{G}_j = \mathbf{F}_{j,0}\mathbf{B}_0 + \sum_{k=1}^{j} \mathbf{E}_{j-k}\mathbf{B}_k \qquad j = 0 \ldots (N_p - 1)$$

Since $\mathbf{F}_{j,0} = \mathbf{E}_j$, the following results:

$$\mathbf{G}_j = \sum_{k=0}^{j} \mathbf{E}_{j-k}\mathbf{B}_k$$

For $k > nb$, the corresponding summands do not exist, since then the terms $\mathbf{B}_k$ do not exist anymore. Therefore, the rows of $\mathbf{G}$ also differ only in one element.

With this (8.15) results to:

$$\begin{bmatrix} \hat{\mathbf{y}}(t+1) \\ \vdots \\ \hat{\mathbf{y}}(t+j) \\ \vdots \\ \hat{\mathbf{y}}(t+N_p) \end{bmatrix} = \begin{bmatrix} \mathbf{G}_0 & \cdots & 0 & \cdots & 0 \\ \vdots & \ddots & \vdots & \ddots & \vdots \\ \mathbf{G}_{j-1} & \cdots & \mathbf{G}_0 & \cdots & 0 \\ \vdots & \ddots & \vdots & \ddots & \vdots \\ \mathbf{G}_{N_p-1} & \cdots & \mathbf{G}_{N_p-j} & \cdots & \mathbf{G}_0 \end{bmatrix} \begin{bmatrix} \Delta\mathbf{u}(t) \\ \vdots \\ \Delta\mathbf{u}(t+j-1) \\ \vdots \\ \Delta\mathbf{u}(t+N_p-1) \end{bmatrix} + \begin{bmatrix} \mathbf{f}(t+1) \\ \vdots \\ \mathbf{f}(t+j) \\ \vdots \\ \mathbf{f}(t+N_p) \end{bmatrix} \tag{8.24}$$

For the calculation of the free response, both matrix polynomials $\mathbf{G}_{jp}(z^{-1})$ and $\mathbf{F}_j(z^{-1})$ are required according to equation (8.12). The calculation of $\mathbf{F}_j(z^{-1})$ has already been explained in the previous chapter. $\mathbf{G}_{jp}(z^{-1})$, however, is calculated exactly like $\mathbf{G}_j(z^{-1})$ via definition (8.11), whereas now, however, the terms with degree $> j$ are of interest, since only past values of the actuating variables are considered in the free response. This is realized via the time shift factor z^{-j} in equation (8.11). If the equation is solved, the individual elements of $\mathbf{G}_{jp}(z^{-1})$ can be obtained:

$$\mathbf{G}_{jp,i} = \sum_{k=0}^{j-1} \mathbf{E}_k \mathbf{B}_{j+i-k} \qquad j = 1 \dots N_p;\ i = 0 \dots (nb-1)$$

Here it has to be noted, too, that summands with $(j+i-k) > nb$ are not considered.

As already done for the SISO system the coefficients of the polynomial matrices $\mathbf{G_p}(z^{-1})$ and $\mathbf{F}(z^{-1})$ are combined so that the following applies:

$$\mathbf{F} = \mathbf{F}(z^{-1})\mathbf{y}(t) + \mathbf{G_p}(z^{-1})\Delta\mathbf{u}(t-1) = \mathbf{FG_p} \cdot \mathbf{YU}$$

in which

$$\mathbf{F}(z^{-1}) = \begin{bmatrix} \mathbf{F}_1(z^{-1}) \\ \vdots \\ \mathbf{F}_{N_p}(z^{-1}) \end{bmatrix} = \begin{bmatrix} \mathbf{F}_{1,0} + \mathbf{F}_{1,1} z^{-1} + \cdots + \mathbf{F}_{1,na} z^{-na} \\ \vdots \\ \mathbf{F}_{N_p,0} + \mathbf{F}_{N_p,1} z^{-1} + \cdots + \mathbf{F}_{N_p,na} z^{-na} \end{bmatrix}$$

$$\mathbf{G_p}(z^{-1}) = \begin{bmatrix} \mathbf{G}_{1p}(z^{-1}) \\ \vdots \\ \mathbf{G}_{N_p p}(z^{-1}) \end{bmatrix} = \begin{bmatrix} \mathbf{G}_{1p,0} + \mathbf{G}_{1p,1} z^{-1} + \cdots + \mathbf{G}_{1p,nb-1} z^{-(nb-1)} \\ \vdots \\ \mathbf{G}_{N_p p,0} + \mathbf{G}_{N_p p,1} z^{-1} + \cdots + \mathbf{G}_{N_p p,nb-1} z^{-(nb-1)} \end{bmatrix}$$

as well as the combinations

$$\mathbf{FG_p} = \begin{bmatrix} \mathbf{F}_{1,0} & \mathbf{F}_{1,1} & \cdots & \mathbf{F}_{1,na} & \mathbf{G}_{1p,0} & \mathbf{G}_{1p,1} & \cdots & \mathbf{G}_{1p,nb-1} \\ \vdots & \vdots & \vdots & \vdots & \vdots & \vdots & \vdots & \vdots \\ \mathbf{F}_{N_p,0} & \mathbf{F}_{N_p,1} & \cdots & \mathbf{F}_{N_p,na} & \mathbf{G}_{N_p p,0} & \mathbf{G}_{N_p p,1} & \cdots & \mathbf{G}_{N_p p,nb-1} \end{bmatrix}$$

$$\mathbf{YU} = \begin{bmatrix} \mathbf{y}(t) \\ \vdots \\ \mathbf{y}(t-na) \\ \Delta\mathbf{u}(t-1) \\ \vdots \\ \Delta\mathbf{u}(t-nb) \end{bmatrix}$$

8.1.4 Consideration of the control horizon

If the actuating variable $\mathbf{u}(t)$ is assumed to be constant after N_u time steps, N_u is called control horizon. Since the actuating variable does not change anymore ($\Delta\mathbf{u}(t) = 0$) for $N_u < t \leqq N_2$, computation time can be saved as full prediction matrices are not required anymore. Thus, equation (8.24), assuming that $N_u \leqq N_2$, simplifies to:

$$\begin{bmatrix} \hat{\mathbf{y}}(t+N_1) \\ \vdots \\ \hat{\mathbf{y}}(t+N_2) \end{bmatrix} = \begin{bmatrix} \mathbf{G}_{N_1-1} & \mathbf{G}_{N_1-2} & \cdots & \mathbf{G}_{N_1-N_u} \\ \vdots & \vdots & & \vdots \\ \mathbf{G}_{N_2-1} & \mathbf{G}_{N_2-2} & \cdots & \mathbf{G}_{N_2-N_u} \end{bmatrix} \begin{bmatrix} \Delta\mathbf{u}(t) \\ \vdots \\ \Delta\mathbf{u}(t+N_u-1) \end{bmatrix} + \begin{bmatrix} \mathbf{f}(t+N_1) \\ \vdots \\ \mathbf{f}(t+N_2) \end{bmatrix}$$

or

$$\mathbf{Y}_{N_{12}} = \mathbf{G}_{N_{12u}} \tilde{\mathbf{U}}_{N_u} + \mathbf{F}_{N_{12}} \tag{8.25}$$

Hence, the equations for the calculation of optimum values for the actuating variables given in chapter 8.1.3 on page 88 can be further simplified. Substituting equation (8.25) instead of (8.15) into (8.14) results in the following cost function:

$$J = (\mathbf{G}_{N_{12u}} \tilde{\mathbf{U}}_{N_u} + \mathbf{F}_{N_{12}} - \mathbf{W}_{N_{12}})^T (\mathbf{G}_{N_{12u}} \tilde{\mathbf{U}}_{N_u} + \mathbf{F}_{N_{12}} - \mathbf{W}_{N_{12}}) + \lambda \tilde{\mathbf{U}}_{N_u}{}^T \tilde{\mathbf{U}}_{N_u}$$

If the minimum of the above equation is determined, optimum values for the actuating variables which take the control horizon N_u into account can be obtained from the following equation:

$$\tilde{\mathbf{U}}_{N_u} = (\mathbf{G}_{N_{12u}}{}^T \mathbf{G}_{N_{12u}} + \lambda\mathbf{I})^{-1} \cdot \mathbf{G}_{N_{12u}}{}^T (\mathbf{W}_{N_{12}} - \mathbf{F}_{N_{12}}) \tag{8.26}$$

Here of course only $\Delta\mathbf{u}(t)$ is required, too. Hence, only the first m rows of $\tilde{\mathbf{U}}$ need to be calculated.

Considering which ones of the equations from page 90 on, which are necessary for the calculation of the free and the forced response, can be simplified in the same way, it can easily be seen that because of the restriction of $\mathbf{F}$ to $\mathbf{F}_{N_{12}}$ not the complete matrix $\mathbf{FG_p}$ is needed. Since, however, $\mathbf{YU}$ contains only the *past* values of $\mathbf{y}(t)$ and $\Delta\mathbf{u}(t)$, no simplification results for this matrix. Analogously, the number of *columns* of $\mathbf{FG_p}_{N_{12}}$ and $\mathbf{FG_p}$ is identical, only the number of *rows* is reduced. Therefore the equation

$$\mathbf{F}_{N_{12}} = \mathbf{FG_p}_{N_{12}} \cdot \mathbf{YU}$$

results, in which

$$\mathbf{FG}_{\mathbf{p}N_{12}} = \begin{bmatrix} \mathbf{F}_{N_1,0} & \cdots & \mathbf{F}_{N_1,na} & \mathbf{G}_{N_1p,0} & \cdots & \mathbf{G}_{N_1p,nb-1} \\ \vdots & \vdots & \vdots & \vdots & \vdots & \vdots \\ \mathbf{F}_{N_2,0} & \cdots & \mathbf{F}_{N_2,na} & \mathbf{G}_{N_2p,0} & \cdots & \mathbf{G}_{N_2p,nb-1} \end{bmatrix}$$

8.2 Consideration of disturbance inputs with GPC

If MPC is used for the control of a nonlinear system, e. g. an induction machine, the nonlinearities of the controlled system can be considered in two different ways. Either the nonlinear parts of the system are approximated with linear ones or a nonlinear variant of MPC can be used. Indeed, until now nonlinear MPC methods still represent a widely uninvestigated field. Certainly, there are several publications which deal with nonlinear MPC control or which introduce such control methods; nevertheless, this technology is still in its fledgling stages. Although the statement by García, Prett and Morari from 1989 [40] that in this research area barely the basics are understood is outdated now since experimental investigations and sporadically even industrial applications of nonlinear MPC methods are demonstrated in newer publications. However, a theoretical foundation, in contrast to the linear strategies, is nearly completely absent until today [88, 102]. Although meanwhile progress was made, nonlinear model-based methods must become much more reliable, more efficient and failsafe before they will be widely accepted. One of the major problems of all nonlinear MPC control schemes is the proof that the control method always finds the global minimum which is not as trivial as in the linear case and partly proofs do still have to be deduced. In the case of nonlinear controls, no conclusions can be drawn from considering the open and then the closed control loop, since the principle of superposition does only apply for linear systems. Unfortunately, it would go beyond the scope of this work to derive a nonlinear optimization method including a nonlinear model and a stability proof. However, as a continuation of this project the development of a nonlinear MPC for electric drives could be conceivable; approaches therefore can be found e. g. in the publication of Morari/Lee [88].

If a drive with an induction machine has to be controlled it makes sense to consider the multiplicatively linked cross couplings which cannot be included as non-linearities in the linear CARIMA model as disturbances instead of implementing a nonlinear control strategy.

8.2.1 Determination of the transfer function

Based on the discrete-time state space representation considering the disturbances according to the equations (7.41) and (7.42) now analog to the transfer function matrix $\mathbf{G_d}(z^{-1})$ in equation (8.1), a disturbance transfer function matrix $\mathbf{H_d}(z^{-1})$ is defined:

$$\mathbf{H_d}(z^{-1}) = \mathbf{H_{d0}} + \mathbf{H_{d1}}z^{-1} + \mathbf{H_{d2}}z^{-2} + \cdots + \mathbf{H_{dn}}z^{-n} = \frac{\mathbf{Y}(z)}{\mathbf{V}(z)} \tag{8.27}$$

whereas the individual elements have the following dimensions:

$$\begin{aligned} \mathbf{y}(t) &: \text{Vector}(n \times 1) \qquad && \text{with } n = \text{number of system outputs} \\ \mathbf{v}(t) &: \text{Vector}(l \times 1) && \text{with } l = \text{number of disturbance inputs} \\ \mathbf{H_d}(z^{-1}) &: \text{Matrix}(n \times l) && \end{aligned}$$

With the help of the approach for the transfer function matrix $\mathbf{G_d}(z^{-1})$, described in chapter 8.1.1, the disturbance transfer function matrix $\mathbf{H_d}(z^{-1})$ can be obtained in the same way:

$$\mathbf{H_d}(z^{-1}) = \mathbf{C_d}(z\mathbf{I} - \mathbf{A_d})^{-1}\mathbf{E_d} + \mathbf{F_d} \tag{8.28}$$

Thus, two mathematically separated systems with overlapping outputs exist for the actuating and for the disturbance variables.

8.2.2 Calculation of the system matrices

If the statements for the calculation of the polynomial matrices $\mathbf{A}(z^{-1})$ and $\mathbf{B}(z^{-1})$, described in chapter 8.1.2, are applied to the matrix $\mathbf{D}(z^{-1})$ which is necessary for the consideration of disturbances in the CARIMA model, the following must be effective:

$$\mathbf{H_d}(z^{-1}) = \mathbf{A}(z^{-1})^{-1}\mathbf{D}(z^{-1})z^{-1}$$

Since $\mathbf{A}(z^{-1})$ is already known from 8.1.2, $\mathbf{D}(z^{-1})$ can be calculated directly with the help of

$$\mathbf{D}(z^{-1}) = \mathbf{A}(z^{-1})\mathbf{H_d}(z^{-1})z \tag{8.29}$$

8.2.3 Mathematical derivation

The CARIMA model

The multidimensional CARIMA model in equation (8.7) is complemented with a deterministic disturbance term so that disturbance variables can be considered:

$$\mathbf{A}(z^{-1})\mathbf{y}(t) = \mathbf{B}(z^{-1})\mathbf{u}(t-1) + \mathbf{D}(z^{-1})\mathbf{v}(t) + \mathbf{C}(z^{-1})\frac{\boldsymbol{\xi}(t)}{\Delta} \tag{8.30}$$

with

$$\mathbf{D}(z^{-1}) = \mathbf{D}_0 + \mathbf{D}_1 z^{-1} + \mathbf{D}_2 z^{-2} + \cdots + \mathbf{D}_{nd} z^{-nd}$$

The dimensions of $\mathbf{D}(z^{-1})$ and $\mathbf{v}(t)$ are:

$$\begin{aligned}\mathbf{D}(z^{-1}) &: \text{Matrix}(n \times l)\\ \mathbf{v}(t) &: \text{Vector}(l \times 1)\end{aligned}$$

Like in the previous cases, it is also assumed here that $\boldsymbol{\xi}(t)$ represents white noise; therefore $\mathbf{C}(z^{-1})$ is equal to the identity matrix $\mathbf{I}$. Thus equation (8.30) becomes to

$$\mathbf{A}(z^{-1})\mathbf{y}(t) = \mathbf{B}(z^{-1})\mathbf{u}(t-1) + \mathbf{D}(z^{-1})\mathbf{v}(t) + \frac{\boldsymbol{\xi}(t)}{\Delta} \tag{8.31}$$

The j-step ahead predictor

If the disturbance variables are considered, the well-known Diophantine equation does not change in comparison to the classical variant. The precalculation of future system outputs can here also be derived via equation (8.9). Multiplicating equation (8.31) with $\Delta\mathbf{E}_j(z^{-1})z^j$ and applying equation (8.9) solved for $\mathbf{E}_j(z^{-1})\tilde{\mathbf{A}}(z^{-1})$, the following results for the prediction:

$$\begin{aligned}\mathbf{y}(t+j) = {} & \mathbf{E}_j(z^{-1})\mathbf{B}(z^{-1})\Delta\mathbf{u}(t+j-1)\\ & + \mathbf{E}_j(z^{-1})\mathbf{D}(z^{-1})\Delta\mathbf{v}(t+j)\\ & + \mathbf{F}_j(z^{-1})\mathbf{y}(t)\\ & + \mathbf{E}_j(z^{-1})\boldsymbol{\xi}(t+j)\end{aligned}$$

Again, the noise terms being in the future can be neglected and thus

$$\begin{aligned}\hat{\mathbf{y}}(t+j) = {} & \mathbf{E}_j(z^{-1})\mathbf{B}(z^{-1})\Delta\mathbf{u}(t+j-1)\\ & + \mathbf{E}_j(z^{-1})\mathbf{D}(z^{-1})\Delta\mathbf{v}(t+j)\\ & + \mathbf{F}_j(z^{-1})\mathbf{y}(t)\end{aligned} \tag{8.32}$$

can be obtained. Now, analog to (8.11)

$$\mathbf{E}_j(z^{-1})\mathbf{D}(z^{-1}) = \mathbf{H}_j(z^{-1}) + z^{-j}\mathbf{H}_{jp}(z^{-1}) \qquad \text{degree}\big(\mathbf{H}_j(z^{-1})\big) < j \tag{8.33}$$

is defined additionally. Since the matrix polynomials $\mathbf{H}_j(z^{-1})$ and $\mathbf{H}_{jp}(z^{-1})$ for the disturbances exactly correspond to $\mathbf{G}_j(z^{-1})$ and $\mathbf{G}_{jp}(z^{-1})$, the statements in chapter 8.1.3 on page 86 et seqq. concerning the decomposition in future and past values apply here, too. Therefore, equation (8.32) becomes

$$\begin{aligned}\hat{\mathbf{y}}(t+j) = {} & \underbrace{\mathbf{G}_j(z^{-1})\Delta\mathbf{u}(t+j-1)}_{\text{forced response}} + \underbrace{\mathbf{H}_j(z^{-1})\Delta\mathbf{v}(t+j)}_{\substack{\text{future, deterministic}\\ \text{disturbances}}} \\ & + \underbrace{\mathbf{G}_{jp}(z^{-1})\Delta\mathbf{u}(t-1) + \mathbf{H}_{jp}(z^{-1})\Delta\mathbf{v}(t) + \mathbf{F}_j(z^{-1})\mathbf{y}(t)}_{\substack{\text{all values in the past}\\ \rightarrow \text{free response } \mathbf{f}(t+j)}}\end{aligned} \tag{8.34}$$

$$\hat{\mathbf{y}}(t+j) = \mathbf{G}_j(z^{-1})\Delta\mathbf{u}(t+j-1) + \mathbf{H}_j(z^{-1})\Delta\mathbf{v}(t+j) + \mathbf{f}(t+j) \tag{8.35}$$

with

$$\mathbf{f}(t+j) = \mathbf{G}_{jp}(z^{-1})\Delta\mathbf{u}(t-1) + \mathbf{H}_{jp}(z^{-1})\Delta\mathbf{v}(t) + \mathbf{F}_j(z^{-1})\mathbf{y}(t)$$

Dimensions:

$$\begin{aligned}\mathbf{G}_j(z^{-1}), \mathbf{G}_{jp}(z^{-1}) &: \text{Matrix}(n \times m) \qquad \text{(unchanged)}\\ \mathbf{H}_j(z^{-1}), \mathbf{H}_{jp}(z^{-1}) &: \text{Matrix}(n \times l)\\ \mathbf{f}(t) &: \text{Vector}(n \times 1)\end{aligned}$$

The individual prediction steps are

$$\begin{aligned}\hat{\mathbf{y}}(t+1) &= \mathbf{G}_1(z^{-1})\Delta\mathbf{u}(t) + \mathbf{H}_1(z^{-1})\Delta\mathbf{v}(t+1) + \mathbf{f}(t+1)\\ \hat{\mathbf{y}}(t+2) &= \mathbf{G}_2(z^{-1})\Delta\mathbf{u}(t+1) + \mathbf{H}_2(z^{-1})\Delta\mathbf{v}(t+2) + \mathbf{f}(t+2)\\ \hat{\mathbf{y}}(t+3) &= \mathbf{G}_3(z^{-1})\Delta\mathbf{u}(t+2) + \mathbf{H}_3(z^{-1})\Delta\mathbf{v}(t+3) + \mathbf{f}(t+3)\\ &\;\;\vdots\\ \hat{\mathbf{y}}(t+N_p) &= \mathbf{G}_{N_p}(z^{-1})\Delta\mathbf{u}(t+N_p-1) + \mathbf{H}_{N_p}(z^{-1})\Delta\mathbf{v}(t+N_p) + \mathbf{f}(t+N_p)\end{aligned}$$

Calculation of the actuating variables

Since nothing has changed in the mathematical basis for the calculation of optimum values for the actuating variables, which is still done via a quadratic cost function, equation (8.14) remains valid without any changes. After a combination of equation (8.35) analog to the procedure in chapter 8.1.3 on page 88, the prediction equation results in

$$\mathbf{Y} = \mathbf{G}\tilde{\mathbf{U}} + \mathbf{H}\tilde{\mathbf{V}} + \mathbf{F} \tag{8.36}$$

with $\tilde{\mathbf{U}} = \Delta\mathbf{U}$ and $\tilde{\mathbf{V}} = \Delta\mathbf{V}$. Since the disturbance term $\mathbf{H}\tilde{\mathbf{V}}$ is independent of the actuating variables, it can be considered as a kind of free response term. Therefore, $\mathbf{H}\tilde{\mathbf{V}}$ and $\mathbf{F}$ are combined to a new free response $\mathbf{F}' = \mathbf{H}\tilde{\mathbf{V}} + \mathbf{F}$. Thus, a simplified prediction equation results:

$$\mathbf{Y} = \mathbf{G}\tilde{\mathbf{U}} + \mathbf{F}' \tag{8.37}$$

Taking these assumptions into account, optimum future values for the actuating variables can be calculated in the same way as without consideration of known disturbances, only the free response $\mathbf{F}$ used there has to be replaced with its modified version $\mathbf{F}'$. The optimum controller outputs can be calculated with

$$\tilde{\mathbf{U}} = (\mathbf{G}^T\mathbf{G} + \lambda\mathbf{I})^{-1} \cdot \mathbf{G}^T(\mathbf{W} - \mathbf{F}') \tag{8.38}$$

if the usual simplifications, $\mu_j = 1$ and $\lambda_j = \lambda$ for all j, are applied.

Recursion of the Diophantine equation

The Diophantine equation which is necessary for the determination of the polynomial matrices for the prediction has not changed compared to MIMO-GPC without consideration of disturbances. Equation (8.9) is valid without any modifications and hence also the approach described in chapter 8.1.3 on page 89 et seq.

Calculation of the free and forced response

Comparing equation (8.34) and (8.35) to (8.12) and (8.13) it can be seen that the polynomial matrices $\mathbf{G}_j(z^{-1})$ and $\mathbf{G}_{jp}(z^{-1})$ can be taken without any changes. Hence, only $\mathbf{H}_j(z^{-1})$ and $\mathbf{H}_{jp}(z^{-1})$ have to be derived newly. Since the definitions of the polynomial matrices mentioned above in equation (8.33) and (8.11) are very similar to each other, $\mathbf{H}_j(z^{-1})$ has to be similar to $\mathbf{G}_j(z^{-1})$

because of the recursive properties of $\mathbf{E}_j(z^{-1})$. Therefore the following must apply:

$$\mathbf{H}_j(z^{-1}) = \mathbf{H}_0 + \mathbf{H}_1 z^{-1} + \mathbf{H}_2 z^{-2} + \cdots + \mathbf{H}_{j-1} z^{-(j-1)} \qquad j = 1 \ldots N_p$$

with

$$\mathbf{H}_j = \sum_{k=0}^{j} \mathbf{E}_{j-k} \mathbf{D}_k$$

All summands with $k > nd$ are omitted as well. Therefore, equation (8.36) can be written in detail as:

$$\begin{bmatrix} \hat{\mathbf{y}}(t+1) \\ \vdots \\ \hat{\mathbf{y}}(t+j) \\ \vdots \\ \hat{\mathbf{y}}(t+N_p) \end{bmatrix} = \begin{bmatrix} \mathbf{G}_0 & \cdots & 0 & \cdots & 0 \\ \vdots & \ddots & \vdots & \ddots & \vdots \\ \mathbf{G}_{j-1} & \cdots & \mathbf{G}_0 & \cdots & 0 \\ \vdots & \ddots & \vdots & \ddots & \vdots \\ \mathbf{G}_{N_p-1} & \cdots & \mathbf{G}_{N_p-j} & \cdots & \mathbf{G}_0 \end{bmatrix} \begin{bmatrix} \Delta\mathbf{u}(t) \\ \vdots \\ \Delta\mathbf{u}(t+j-1) \\ \vdots \\ \Delta\mathbf{u}(t+N_p-1) \end{bmatrix}$$
$$+ \begin{bmatrix} \mathbf{H}_0 & \cdots & 0 & \cdots & 0 \\ \vdots & \ddots & \vdots & \ddots & \vdots \\ \mathbf{H}_{j-1} & \cdots & \mathbf{H}_0 & \cdots & 0 \\ \vdots & \ddots & \vdots & \ddots & \vdots \\ \mathbf{H}_{N_p-1} & \cdots & \mathbf{H}_{N_p-j} & \cdots & \mathbf{H}_0 \end{bmatrix} \begin{bmatrix} \Delta\mathbf{v}(t+1) \\ \vdots \\ \Delta\mathbf{v}(t+j) \\ \vdots \\ \Delta\mathbf{v}(t+N_p) \end{bmatrix}$$
$$+ \begin{bmatrix} \mathbf{f}(t+1) \\ \vdots \\ \mathbf{f}(t+j) \\ \vdots \\ \mathbf{f}(t+N_p) \end{bmatrix} \tag{8.39}$$

Besides the already known terms $\mathbf{G}_{jp}(z^{-1})$ and $\mathbf{F}_j(z^{-1})$ after equation (8.34), the term $\mathbf{H}_{jp}(z^{-1})$ is also needed for the calculation of the free response. Since this term is also calculated according to (8.33) analog to the calculation of $\mathbf{G}_{jp}(z^{-1})$, the following applies for the individual elements of $\mathbf{H}_{jp}(z^{-1})$:

$$\mathbf{H}_{jp,i} = \sum_{k=0}^{j-1} \mathbf{E}_k \mathbf{D}_{j+i-k} \qquad j = 1 \ldots N_p;\ i = 0 \ldots (nd-1)$$

Summands with $(j+i-k) > nd$ are omitted again.

If the coefficients of the polynomial matrices $\mathbf{G_p}(z^{-1})$, $\mathbf{H_p}(z^{-1})$ and $\mathbf{F}(z^{-1})$ are combined together into a single matrix this results in

$$\mathbf{F} = \mathbf{F}(z^{-1})\mathbf{y}(t) + \mathbf{G_p}(z^{-1})\Delta\mathbf{u}(t-1) + \mathbf{H_p}(z^{-1})\Delta\mathbf{v}(t) = \mathbf{FG_pH_p} \cdot \mathbf{YUV}$$

in which, in addition to the already known definitions $\mathbf{F}(z^{-1})$ and $\mathbf{G}(z^{-1})$, the terms

$$\mathbf{H_p}(z^{-1}) = \begin{bmatrix} \mathbf{H}_{1p}(z^{-1}) \\ \vdots \\ \mathbf{H}_{N_pp}(z^{-1}) \end{bmatrix} = \begin{bmatrix} \mathbf{H}_{1p,0} + \mathbf{H}_{1p,1}z^{-1} + \cdots + \mathbf{H}_{1p,nd-1}z^{-(nd-1)} \\ \vdots \\ \mathbf{H}_{N_pp,0} + \mathbf{H}_{N_pp,1}z^{-1} + \cdots + \mathbf{H}_{N_pp,nd-1}z^{-(nd-1)} \end{bmatrix}$$

$$\mathbf{FG_pH_p} = \begin{bmatrix} \mathbf{F}_{1,0} & \mathbf{F}_{1,1} & \cdots & \mathbf{F}_{1,na} \\ \vdots & \vdots & \vdots & \vdots \\ \mathbf{F}_{N_p,0} & \mathbf{F}_{N_p,1} & \cdots & \mathbf{F}_{N_p,na} \\ \mathbf{G}_{1p,0} & \mathbf{G}_{1p,1} & \cdots & \mathbf{G}_{1p,nb-1} \\ \vdots & \vdots & \vdots & \vdots \\ \mathbf{G}_{N_pp,0} & \mathbf{G}_{N_pp,1} & \cdots & \mathbf{G}_{N_pp,nb-1} \\ \mathbf{H}_{1p,0} & \mathbf{H}_{1p,1} & \cdots & \mathbf{H}_{1p,nd-1} \\ \vdots & \vdots & \vdots & \vdots \\ \mathbf{H}_{N_pp,0} & \mathbf{H}_{N_pp,1} & \cdots & \mathbf{H}_{N_pp,nd-1} \end{bmatrix}$$

$$\mathbf{YUV} = \begin{bmatrix} \mathbf{y}(t) \\ \vdots \\ \mathbf{y}(t-na) \\ \Delta\mathbf{u}(t-1) \\ \vdots \\ \Delta\mathbf{u}(t-nb) \\ \Delta\mathbf{v}(t) \\ \vdots \\ \Delta\mathbf{v}(t-nd+1) \end{bmatrix}$$

are valid. For a further simplification of the matrix operations, $\mathbf{F}' = \mathbf{H}\tilde{\mathbf{V}} + \mathbf{F}$ can be summarized again as already explained. However, this does only make sense if the future values of the disturbances $\mathbf{v}(t)$ are known. If they are unknown, it is more reasonable to choose the actual value for all future values, i. e. $\mathbf{v}(t+j) = \mathbf{v}(t)$. Thereby, the changes $\Delta\mathbf{v}(t+j)$ become zero which leads to the fact that the complete term $\mathbf{H}\tilde{\mathbf{V}}$ disappears. In these cases, the calculation of a modified free response $\mathbf{F}'$ is unnecessary; simply $\mathbf{F}$ is used instead of $\mathbf{F}'$ in equation (8.38). However, if the future disturbances are explicitly known, $\mathbf{F}'$ should be determined exactly. Then the modified free response results to

$$\mathbf{F}' = \mathbf{H} \cdot \tilde{\mathbf{V}} + \mathbf{FG_pH_p} \cdot \mathbf{YUV} = \mathbf{HFG_pH_p} \cdot \tilde{\mathbf{V}}\mathbf{YUV}$$

in which

$$\mathbf{HFG_pH_p} = \begin{bmatrix} \mathbf{H}_0 & 0 & \cdots & 0 \\ \mathbf{H}_1 & \mathbf{H}_0 & \cdots & 0 \\ \vdots & \vdots & \ddots & \vdots \\ \mathbf{H}_{N_p-1} & \mathbf{H}_{N_p-2} & \cdots & \mathbf{H}_0 \\ \mathbf{F}_{1,0} & \mathbf{F}_{1,1} & \cdots & \mathbf{F}_{1,na} \\ \mathbf{F}_{2,0} & \mathbf{F}_{2,1} & \cdots & \mathbf{F}_{2,na} \\ \vdots & \vdots & \vdots & \vdots \\ \mathbf{F}_{N_p,0} & \mathbf{F}_{N_p,1} & \cdots & \mathbf{F}_{N_p,na} \\ \mathbf{G}_{1p,0} & \mathbf{G}_{1p,1} & \cdots & \mathbf{G}_{1p,nb-1} \\ \mathbf{G}_{2p,0} & \mathbf{G}_{2p,1} & \cdots & \mathbf{G}_{2p,nb-1} \\ \vdots & \vdots & \vdots & \vdots \\ \mathbf{G}_{N_pp,0} & \mathbf{G}_{N_pp,1} & \cdots & \mathbf{G}_{N_pp,nb-1} \\ \mathbf{H}_{1p,0} & \mathbf{H}_{1p,1} & \cdots & \mathbf{H}_{1p,nd-1} \\ \mathbf{H}_{2p,0} & \mathbf{H}_{2p,1} & \cdots & \mathbf{H}_{2p,nd-1} \\ \vdots & \vdots & \vdots & \vdots \\ \mathbf{H}_{N_pp,0} & \mathbf{H}_{N_pp,1} & \cdots & \mathbf{H}_{N_pp,nd-1} \end{bmatrix}$$

$$
\tilde{\mathbf{V}}\mathbf{YUV} = \begin{bmatrix} \Delta\mathbf{v}(t+1) \\ \vdots \\ \Delta\mathbf{v}(t+N_p) \\ \mathbf{y}(t) \\ \vdots \\ \mathbf{y}(t-na) \\ \Delta\mathbf{u}(t-1) \\ \vdots \\ \Delta\mathbf{u}(t-nb) \\ \Delta\mathbf{v}(t) \\ \vdots \\ \Delta\mathbf{v}(t-nd+1) \end{bmatrix}
$$

8.2.4 Consideration of the control horizon

Analog to the approach described in chapter 8.1.4, the computation time necessary for the calculation of optimum future values for the actuating variables should also be reduced when GPC with consideration of disturbances is used by taking the control horizon N_u with $N_u \leqq N_2$ into account. The parts of the matrices $\mathbf{G}$, $\mathbf{H}$, $\mathbf{F}'$ etc., which are unnecessary for the calculation of $\Delta\mathbf{u}(t+j)$, $j - N_1 \ldots N_2$ are neglected and the matrices are reduced in their dimensions to sizes as small as possible. A prediction equation with lesser dimensions results:

$$
\begin{bmatrix} \hat{\mathbf{y}}(t+N_1) \\ \vdots \\ \hat{\mathbf{y}}(t+N_2) \end{bmatrix} = \begin{bmatrix} \mathbf{G}_{N_1-1} & \mathbf{G}_{N_1-2} & \cdots & \mathbf{G}_{N_1-N_u} \\ \vdots & \vdots & & \vdots \\ \mathbf{G}_{N_2-1} & \mathbf{G}_{N_2-2} & \cdots & \mathbf{G}_{N_2-N_u} \end{bmatrix} \begin{bmatrix} \Delta\mathbf{u}(t) \\ \vdots \\ \Delta\mathbf{u}(t+N_u-1) \end{bmatrix}
+ \begin{bmatrix} \mathbf{H}_{N_1-1} & \mathbf{H}_{N_1-2} & \cdots & \mathbf{H}_{N_1-N_u} \\ \vdots & \vdots & & \vdots \\ \mathbf{H}_{N_2-1} & \mathbf{H}_{N_2-2} & \cdots & \mathbf{H}_{N_2-N_u} \end{bmatrix} \begin{bmatrix} \Delta\mathbf{v}(t+1) \\ \vdots \\ \Delta\mathbf{v}(t+N_u) \end{bmatrix}
+ \begin{bmatrix} \mathbf{f}(t+N_1) \\ \vdots \\ \mathbf{f}(t+N_2) \end{bmatrix}
$$

or

$$\mathbf{Y}_{N_{12}} = \mathbf{G}_{N_{12u}}\tilde{\mathbf{U}}_{N_u} + \mathbf{H}_{N_{12u}}\tilde{\mathbf{V}} + \mathbf{F}_{N_{12}} \tag{8.40}$$

In the same way, equation (8.38) for the calculation of optimum values for the actuating variables can be simplified to

$$\tilde{\mathbf{U}}_{N_u} = (\mathbf{G}_{N_{12u}}{}^T\mathbf{G}_{N_{12u}} + \lambda\mathbf{I})^{-1} \cdot \mathbf{G}_{N_{12u}}{}^T(\mathbf{W}_{N_{12}} - \mathbf{F'}_{N_{12}}) \tag{8.41}$$

Since in most cases, especially in the application for a drive control that is described here, future disturbances will not be known, the calculation of a modified free response $\mathbf{F'}_{N_{12}}$ and with it of a transfer function matrix $\mathbf{H}_{N_{12u}}$ is unnecessary. Therefore, for the prediction of the free response, only the equation

$$\mathbf{F}_{N_{12}} = \mathbf{FG_pH_p}_{N_{12}} \cdot \mathbf{YUV}$$

is required, in which

$$\mathbf{FG_pH_p}_{N_{12}} = \begin{bmatrix} \mathbf{F}_{N_1,0} & \mathbf{F}_{N_1,1} & \cdots & \mathbf{F}_{N_1,na} \\ \vdots & \vdots & \vdots & \vdots \\ \mathbf{F}_{N_2,0} & \mathbf{F}_{N_2,1} & \cdots & \mathbf{F}_{N_2,na} \\ \mathbf{G}_{N_1p,0} & \mathbf{G}_{N_1p,1} & \cdots & \mathbf{G}_{N_1p,nb-1} \\ \vdots & \vdots & \vdots & \vdots \\ \mathbf{G}_{N_2p,0} & \mathbf{G}_{N_2p,1} & \cdots & \mathbf{G}_{N_2p,nb-1} \\ \mathbf{H}_{N_1p,0} & \mathbf{H}_{N_1p,1} & \cdots & \mathbf{H}_{N_1p,nd-1} \\ \vdots & \vdots & \vdots & \vdots \\ \mathbf{H}_{N_2p,0} & \mathbf{H}_{N_2p,1} & \cdots & \mathbf{H}_{N_2p,nd-1} \end{bmatrix}$$

8.3 MIMO-GPC with filter

As already described for unidimensional model-based controllers (chapter 6), a MIMO-GPC controller is also not really feasible in the presence of harmonic affected measured values. The possible workarounds, i. e. stabilizing the control loop by reducing the controller dynamics or by integrating a low-pass filter into the measuring system, can absolutely not be recommended as they lead to an all in all more sluggish control. A remedial approach for this, as already described in chapter 6.2 for the unidimensional case, is to integrate a filter into the GPC controller itself. In the following, this should be adapted for a multidimensional control.

Of course, the use of the integrated filter in the system model is possible with and also without consideration of the disturbances. For reasons of simplicity, here only the procedure with consideration of known disturbances is derived. The reader himself can easily derive a GPC control with filter, but without considering the disturbances; in this case only the terms $\mathbf{D}(z^{-1})$ and $\mathbf{v}(t)$ as well as the terms derived from these have to be neglected in the following description.

For a practical implementation of the filter characteristics, Camacho/Bordons [20, chapter 6.1.2] describe a method that is taken from Goodwin/Sin [44, chapter 7.4.2]; however, it seems to be more easy and more reasonable to extend the principle discussed in chapter 6.2.1 on page 50 et seqq. to multidimensional control with consideration of disturbances similar to the SISO principle by Clarke [28].

8.3.1 Determination of the transfer function

Since the integrated filter has no influence on the system itself, the transfer function of the system does of course not change. The equations derived in the chapters 8.1.1 and 8.2.1 apply without any limitations.

8.3.2 Calculation of the system matrices

Being a decomposition of the transfer function, the system matrices do also not depend on the filtering. Hence, the equations specified in the chapters 8.1.2 and 8.2.2 can be applied further on.

8.3.3 Mathematical derivation

The CARIMA model

Since the plant itself has not changed, the CARIMA model (8.30) from chapter 8.2.3 has not changed, either; only the assumption that $\boldsymbol{\xi}(t)$ represents white noise cannot be accepted anymore. Therefore, as already described for the unidimensional case in chapter 6.2.1 on page 50, $\mathbf{C}(z^{-1})$ must not be set equal to the identity matrix $\mathbf{I}$, but is substituted with a design polynomial

$\mathbf{T}(z^{-1})$. Thus, the CARIMA model results into

$$\mathbf{A}(z^{-1})\mathbf{y}(t) = \mathbf{B}(z^{-1})\mathbf{u}(t-1) + \mathbf{D}(z^{-1})\mathbf{v}(t) + \mathbf{T}(z^{-1})\frac{\boldsymbol{\xi}(t)}{\Delta} \tag{8.42}$$

with

$$\mathbf{T}(z^{-1}) = \mathbf{T}_0 + \mathbf{T}_1 z^{-1} + \mathbf{T}_2 z^{-2} + \cdots + \mathbf{T}_{nt} z^{-nt}$$

in which for a filter, normally $\mathbf{T}_0 = \mathbf{I}$ is set. The dimension of $\mathbf{T}(z^{-1})$ is

$$\mathbf{T}(z^{-1}) : \text{Matrix}(n \times n)$$

The j-step ahead predictor

Compared to equation (8.31), an additional term with $\mathbf{T}(z^{-1})$ exists now and so the Diophantine equation (8.9) changes to

$$\begin{aligned}\mathbf{T}(z^{-1}) &= \mathbf{E}_j(z^{-1})\mathbf{A}(z^{-1})\Delta + z^{-j}\mathbf{F}_j(z^{-1}) \\ &= \mathbf{E}_j(z^{-1})\tilde{\mathbf{A}}(z^{-1}) + z^{-j}\mathbf{F}_j(z^{-1})\end{aligned} \tag{8.43}$$

Analog to the approach described in chapter 8.1.3 on page 86 et seqq. by multiplying (8.42) with $\Delta\mathbf{E}_j(z^{-1})z^j$ and by substituting equation (8.43) solved for $\mathbf{E}_j(z^{-1})\tilde{\mathbf{A}}(z^{-1})$ into the result, the following prediction equation can be obtained:

$$\begin{aligned}\mathbf{T}(z^{-1})\mathbf{y}(t+j) = &\ \mathbf{E}_j(z^{-1})\mathbf{B}(z^{-1})\Delta\mathbf{u}(t+j-1) \\ &+ \mathbf{E}_j(z^{-1})\mathbf{D}(z^{-1})\Delta\mathbf{v}(t+j) \\ &+ \mathbf{F}_j(z^{-1})\mathbf{y}(t) \\ &+ \mathbf{T}(z^{-1})\mathbf{E}_j(z^{-1})\boldsymbol{\xi}(t+j)\end{aligned}$$

Again, as $\mathbf{E}_j(z^{-1})$ is of degree $j-1$, all noise terms are completely in the future, which means that they do not have to be considered for the prediction. At the same time, all variables can be filtered with $\mathbf{T}(z^{-1})$, i. e. dividing them by $\mathbf{T}(z^{-1})$. Then, the following results:

$$\begin{aligned}\hat{\mathbf{y}}(t+j) = &\ \mathbf{E}_j(z^{-1})\mathbf{B}(z^{-1})\Delta\mathbf{u}^f(t+j-1) \\ &+ \mathbf{E}_j(z^{-1})\mathbf{D}(z^{-1})\Delta\mathbf{v}^f(t+j) \\ &+ \mathbf{F}_j(z^{-1})\mathbf{y}^f(t)\end{aligned}$$

with the filtered variables

$$\mathbf{u}^f(t) = \frac{\mathbf{u}(t)}{\mathbf{T}(z^{-1})}\,, \qquad \mathbf{v}^f(t) = \frac{\mathbf{v}(t)}{\mathbf{T}(z^{-1})}\,, \qquad \mathbf{y}^f(t) = \frac{\mathbf{y}(t)}{\mathbf{T}(z^{-1})}$$

Future values of the actuating variables are not known yet. Hence, it does not make sense to filter them; the same applies for future disturbances. Therefore, the polynomial $\mathbf{T}(z^{-1})$ has to be taken out of the corresponding terms. Analog to the equations (6.19), (8.11) and (8.33), the following definitions result:

$$\mathbf{E}_j(z^{-1})\mathbf{B}(z^{-1}) = \mathbf{G}_j{}'(z^{-1})\mathbf{T}(z^{-1}) + z^{-j}\mathbf{\Gamma}_j(z^{-1}) \tag{8.44}$$

$$\mathbf{E}_j(z^{-1})\mathbf{D}(z^{-1}) = \mathbf{H}_j{}'(z^{-1})\mathbf{T}(z^{-1}) + z^{-j}\mathbf{\Theta}_j(z^{-1}) \tag{8.45}$$

With these, the prediction equation becomes

$$\begin{aligned}\hat{\mathbf{y}}(t+j) = {} & \underbrace{\mathbf{G}_j{}'(z^{-1})\Delta\mathbf{u}(t+j-1)}_{\text{forced response}} + \underbrace{\mathbf{H}_j{}'(z^{-1})\Delta\mathbf{v}(t+j)}_{\substack{\text{future deterministic}\\ \text{disturbances}}} \\ & + \underbrace{\mathbf{\Gamma}_j(z^{-1})\Delta\mathbf{u}^f(t-1) + \mathbf{\Theta}_j(z^{-1})\Delta\mathbf{v}^f(t) + \mathbf{F}_j(z^{-1})\mathbf{y}^f(t)}_{\substack{\text{all terms in the past}\\ \to \text{ free response } \mathbf{f}'(t+j)}}\end{aligned} \tag{8.46}$$

$$\hat{\mathbf{y}}(t+j) = \mathbf{G}_j{}'(z^{-1})\Delta\mathbf{u}(t+j-1) + \mathbf{H}_j{}'(z^{-1})\Delta\mathbf{v}(t+j) + \mathbf{f}'(t+j) \tag{8.47}$$

with

$$\mathbf{f}'(t+j) = \mathbf{\Gamma}_j(z^{-1})\Delta\mathbf{u}^f(t-1) + \mathbf{\Theta}_j(z^{-1})\Delta\mathbf{v}^f(t) + \mathbf{F}_j(z^{-1})\mathbf{y}^f(t)$$

Dimensions:

$$\begin{aligned}\mathbf{G}_j{}'(z^{-1}), \mathbf{\Gamma}_j(z^{-1}) &: \text{Matrix}(n \times m) \\ \mathbf{H}_j{}'(z^{-1}), \mathbf{\Theta}_j(z^{-1}) &: \text{Matrix}(n \times l) \\ \mathbf{f}'(t) &: \text{Vector}(n \times 1)\end{aligned}$$

As the equations (8.42) to (8.46) show, the disturbance variables $\mathbf{v}(t)$ have to be filtered with the same filter as the actuating variables $\mathbf{u}(t)$ and the output variables $\mathbf{y}(t)$. Hence, the three vectors mentioned above must have the same dimensions, i. e. $n = m = l$. If this is not the case, the corresponding system matrices of the machine model from chapter 7.3 can be extended as a workaround by artificially adding zero elements in order to get the required number of rows. In this way it can be made sure that all system variables can be filtered with the identical filter matrix.

Finally, the single prediction steps can be written as

$$\hat{\mathbf{y}}(t+1) = \mathbf{G}_1{}'(z^{-1})\Delta\mathbf{u}(t) + \mathbf{H}_1{}'(z^{-1})\Delta\mathbf{v}(t+1) + \mathbf{f}'(t+1)$$

$$\hat{\mathbf{y}}(t+2) = \mathbf{G}_2{}'(z^{-1})\Delta\mathbf{u}(t+1) + \mathbf{H}_2{}'(z^{-1})\Delta\mathbf{v}(t+2) + \mathbf{f}'(t+2)$$

$$\hat{\mathbf{y}}(t+3) = \mathbf{G}_3{}'(z^{-1})\Delta\mathbf{u}(t+2) + \mathbf{H}_3{}'(z^{-1})\Delta\mathbf{v}(t+3) + \mathbf{f}'(t+3)$$

$$\vdots$$

$$\hat{\mathbf{y}}(t+N_p) = \mathbf{G}_{N_p}{}'(z^{-1})\Delta\mathbf{u}(t+N_p-1) + \mathbf{H}_{N_p}{}'(z^{-1})\Delta\mathbf{v}(t+N_p) + \mathbf{f}'(t+N_p)$$

Calculation of the actuating variables

For the determination of optimum future values for the actuating variables, the procedure is just the same as described in chapter 8.1.3 and 8.2.3 on the pages 88 and 99 for MIMO-GPC without internal filter and without or with consideration of disturbances. From equation (8.47) results:

$$\mathbf{Y} = \mathbf{G}'\tilde{\mathbf{U}} + \mathbf{H}'\tilde{\mathbf{V}} + \mathbf{F}' \tag{8.48}$$

Combining the partial terms $\mathbf{H}'\tilde{\mathbf{V}}$ and $\mathbf{F}'$, which are independent of the actuating variables, results in

$$\mathbf{Y} = \mathbf{G}'\tilde{\mathbf{U}} + \mathbf{F}'' \qquad \text{with} \quad \mathbf{F}'' = \mathbf{H}'\tilde{\mathbf{V}} + \mathbf{F}' \tag{8.49}$$

Applying the cost function (8.14), which has already been described in chapter 8.1.3 on page 88, finally, the equation for optimum future values of the actuating variables is obtained:

$$\tilde{\mathbf{U}} = (\mathbf{G}'^T\mathbf{G}' + \lambda\mathbf{I})^{-1} \cdot \mathbf{G}'^T(\mathbf{W} - \mathbf{F}'') \tag{8.50}$$

Recursion of the Diophantine equation

In order to solve the Diophantine equation (8.43) for the multidimensional case, the procedure is exactly the same as the one already described for the unidimensional case in chapter 6.2.1 on page 52. Again, the filter or design polynomial $\mathbf{T}(z^{-1})$ is omitted because of the subtraction. Hence, the recursion equations (8.19) to (8.21) determined in chapter 8.1.3 on page 89 do also apply here. As explained in chapter 8.2.3 on page 99, the consideration of disturbances does not change the Diophantine equation and its solution.

The final terms of the recursion are obtained by considering equation (8.43) for $j = 1$:

$$\mathbf{E}_1(z^{-1}) = \mathbf{E}_0 = \frac{\mathbf{T}_0}{\tilde{\mathbf{A}}_0} = \mathbf{I} \qquad \text{since } \tilde{\mathbf{A}}_0 = \mathbf{I} \text{ and } \mathbf{T}_0 = \mathbf{I} \tag{8.51}$$

$$\mathbf{F}_1(z^{-1}) = z\big(\mathbf{T}(z^{-1}) - \tilde{\mathbf{A}}(z^{-1})\big) \tag{8.52}$$

Calculation of the free and forced response

The individual elements for the calculation of the free and forced response are calculated in the same way as for the unidimensional case. Starting from equation (8.44) with the considerations described in chapter 6.2.1 on page 53 et seqq., one can see that here again no terms with $\mathbf{u}(t+j), j < 0$ may enter $\mathbf{G}_j{}'(z^{-1})$. Because of this, the degree of $\mathbf{G}_j{}'(z^{-1})$ is limited to $\leqq (j-1)$. Accordingly, for the elements of $\mathbf{G}_j{}'(z^{-1})$ follows:

1. The matrix elements $\mathbf{G}'_{j,0} \ldots \mathbf{G}'_{j,j-1}$ and $\mathbf{G}'_{j+1,0} \ldots \mathbf{G}'_{j+1,j-1}$ are all identical.
2. The matrix rows only differ in the newly added element $\mathbf{G}'_{j+1,j}$.

Hence, the following definition applies for $\mathbf{G}'$:

$$\mathbf{G}' = \begin{bmatrix} \mathbf{G}'_0 & 0 & 0 & \cdots & 0 \\ \mathbf{G}'_1 & \mathbf{G}'_0 & 0 & \cdots & 0 \\ \mathbf{G}'_2 & \mathbf{G}'_1 & \mathbf{G}'_0 & \cdots & 0 \\ \vdots & \vdots & \vdots & \ddots & \vdots \\ \mathbf{G}'_{N_p-1} & \mathbf{G}'_{N_p-2} & \mathbf{G}'_{N_p-3} & \cdots & \mathbf{G}'_0 \end{bmatrix}$$

Now, the determination of the elements of $\mathbf{\Gamma}(z^{-1})$ is done with the approach described on page 53 et seqq.: Equation (8.44) is set up for $j+1$ and j and then these equations are subtracted from each other.

$$\begin{array}{r|r} + & \mathbf{E}_{j+1}(z^{-1})\mathbf{B}(z^{-1}) = \mathbf{G}_{j+1}{}'(z^{-1})\mathbf{T}(z^{-1}) + z^{-(j+1)}\mathbf{\Gamma}_{j+1}(z^{-1}) \\ - & \mathbf{E}_j(z^{-1})\mathbf{B}(z^{-1}) = \mathbf{G}_j{}'(z^{-1})\mathbf{T}(z^{-1}) + z^{-j}\mathbf{\Gamma}_j(z^{-1}) \\ \hline & \mathbf{E}_j z^{-j}\mathbf{B}(z^{-1}) = \mathbf{G}'_j z^{-j}\mathbf{T}(z^{-1}) + z^{-j}\big(z^{-1}\mathbf{\Gamma}_{j+1}(z^{-1}) - \mathbf{\Gamma}_j(z^{-1})\big) \end{array}$$

A coefficient comparison results into

$$\mathbf{G}'_j = \frac{\mathbf{E}_j\mathbf{B}_0 + \mathbf{\Gamma}_{j,0}}{\mathbf{T}_0}$$

$$\mathbf{\Gamma}_{j+1,i-1} = \mathbf{\Gamma}_{j,i} + \mathbf{E}_j\mathbf{B}_i - \mathbf{G}'_j\mathbf{T}_i$$

and for the last element:

$$\mathbf{\Gamma}_{j+1,i} = \mathbf{E}_j\mathbf{B}_{i+1} - \mathbf{G'}_j\mathbf{T}_{i+1}$$

The final terms of the recursion can be determined from equation (8.44) for $j = 1$ as follows:

$$\mathbf{G'}_0 = \frac{\mathbf{E}_0\mathbf{B}_0}{\mathbf{T}_0}$$
$$\mathbf{\Gamma}_{1,i} = \mathbf{E}_0\mathbf{B}_{i+1} - \mathbf{G'}_0\mathbf{T}_{i+1}$$

The limits or degrees of the individual polynomial matrices can be derived according to the same criteria as for SISO control with filter. In the same way, all terms with $\mathbf{B}_k, k > nb$ and $\mathbf{T}_k, k > nt$ are omitted again.

Besides $\mathbf{\Gamma}$, $\mathbf{\Theta}$ is also necessary for the calculation of the free response. Since the equations (8.44) and (8.45) have the absolutely identical framework, the same solution approach as for the calculation of $\mathbf{G'}$ and $\mathbf{\Gamma}$ can be used for the calculation of the elements of $\mathbf{H'}$ and $\mathbf{\Theta}$. Then it follows:

$$\mathbf{H'} = \begin{bmatrix} \mathbf{H'}_0 & 0 & 0 & \cdots & 0 \\ \mathbf{H'}_1 & \mathbf{H'}_0 & 0 & \cdots & 0 \\ \mathbf{H'}_2 & \mathbf{H'}_1 & \mathbf{H'}_0 & \cdots & 0 \\ \vdots & \vdots & \vdots & \ddots & \vdots \\ \mathbf{H'}_{N_p-1} & \mathbf{H'}_{N_p-2} & \mathbf{H'}_{N_p-3} & \cdots & \mathbf{H'}_0 \end{bmatrix}$$

The recursion equations are given by:

$$\mathbf{H'}_j = \frac{\mathbf{E}_j\mathbf{D}_0 + \mathbf{\Theta}_{j,0}}{\mathbf{T}_0}$$
$$\mathbf{\Theta}_{j+1,i-1} = \mathbf{\Theta}_{j,i} + \mathbf{E}_j\mathbf{D}_i - \mathbf{H'}_j\mathbf{T}_i$$

and for the last element:

$$\mathbf{\Theta}_{j+1,i} = \mathbf{E}_j\mathbf{D}_{i+1} - \mathbf{H'}_j\mathbf{T}_{i+1}$$

as well as for the last terms:

$$\mathbf{H'}_0 = \frac{\mathbf{E}_0\mathbf{D}_0}{\mathbf{T}_0}$$
$$\mathbf{\Theta}_{1,i} = \mathbf{E}_0\mathbf{D}_{i+1} - \mathbf{H'}_0\mathbf{T}_{i+1}$$

Elements with $\mathbf{D}_k, k > nd$ and $\mathbf{T}_k, k > nt$ are, of course, omitted. Hence, the prediction equation (8.48) can be written in detail as:

$$\begin{bmatrix} \hat{\mathbf{y}}(t+1) \\ \vdots \\ \hat{\mathbf{y}}(t+j) \\ \vdots \\ \hat{\mathbf{y}}(t+N_p) \end{bmatrix} = \begin{bmatrix} \mathbf{G'}_0 & \cdots & 0 & \cdots & 0 \\ \vdots & \ddots & \vdots & \ddots & \vdots \\ \mathbf{G'}_{j-1} & \cdots & \mathbf{G'}_0 & \cdots & 0 \\ \vdots & \ddots & \vdots & \ddots & \vdots \\ \mathbf{G'}_{N_p-1} & \cdots & \mathbf{G'}_{N_p-j} & \cdots & \mathbf{G'}_0 \end{bmatrix} \begin{bmatrix} \Delta\mathbf{u}(t) \\ \vdots \\ \Delta\mathbf{u}(t+j-1) \\ \vdots \\ \Delta\mathbf{u}(t+N_p-1) \end{bmatrix}$$
$$+ \begin{bmatrix} \mathbf{H'}_0 & \cdots & 0 & \cdots & 0 \\ \vdots & \ddots & \vdots & \ddots & \vdots \\ \mathbf{H'}_{j-1} & \cdots & \mathbf{H'}_0 & \cdots & 0 \\ \vdots & \ddots & \vdots & \ddots & \vdots \\ \mathbf{H'}_{N_p-1} & \cdots & \mathbf{H'}_{N_p-j} & \cdots & \mathbf{H'}_0 \end{bmatrix} \begin{bmatrix} \Delta\mathbf{v}(t+1) \\ \vdots \\ \Delta\mathbf{v}(t+j) \\ \vdots \\ \Delta\mathbf{v}(t+N_p) \end{bmatrix}$$
$$+ \begin{bmatrix} \mathbf{f}'(t+1) \\ \vdots \\ \mathbf{f}'(t+j) \\ \vdots \\ \mathbf{f}'(t+N_p) \end{bmatrix} \tag{8.53}$$

For a simpler calculation of the free response $\mathbf{F}'$, the filtered past values $\mathbf{y}^f(t)$, $\Delta\mathbf{u}^f(t)$ and $\Delta\mathbf{v}^f(t)$ are combined to the matrices $\mathbf{Y}^f$, $\mathbf{U}^f$ and $\mathbf{V}^f$ and then they are altogether combined to a big matrix $\mathbf{YUV}^f$ according to the already well-known procedure. In the same way the polynomial matrices $\mathbf{F}(z^{-1})$, $\mathbf{\Gamma}(z^{-1})$ and $\mathbf{\Theta}(z^{-1})$ are combined to $\mathbf{F\Gamma\Theta}$. Then it follows:

$$\mathbf{F}' = \mathbf{F}(z^{-1})\mathbf{y}^f(t) + \mathbf{\Gamma}(z^{-1})\Delta\mathbf{u}^f(t-1) + \mathbf{\Theta}(z^{-1})\Delta\mathbf{v}^f(t) = \mathbf{F\Gamma\Theta} \cdot \mathbf{YUV}^f$$

in which

$$\mathbf{F}(z^{-1}) = \begin{bmatrix} \mathbf{F}_1(z^{-1}) \\ \vdots \\ \mathbf{F}_{N_p}(z^{-1}) \end{bmatrix} = \begin{bmatrix} \mathbf{F}_{1,0} + \mathbf{F}_{1,1}z^{-1} + \cdots + \mathbf{F}_{1,na}z^{-na} \\ \vdots \\ \mathbf{F}_{N_p,0} + \mathbf{F}_{N_p,1}z^{-1} + \cdots + \mathbf{F}_{N_p,na}z^{-na} \end{bmatrix}$$

$$\mathbf{\Gamma}(z^{-1}) = \begin{bmatrix} \mathbf{\Gamma}_1(z^{-1}) \\ \vdots \\ \mathbf{\Gamma}_{N_p}(z^{-1}) \end{bmatrix} = \begin{bmatrix} \mathbf{\Gamma}_{1,0} + \mathbf{\Gamma}_{1,1} z^{-1} + \cdots + \mathbf{\Gamma}_{1,nb-1} z^{-(nb-1)} \\ \vdots \\ \mathbf{\Gamma}_{N_p,0} + \mathbf{\Gamma}_{N_p,1} z^{-1} + \cdots + \mathbf{\Gamma}_{N_p,nb-1} z^{-(nb-1)} \end{bmatrix}$$

$$\mathbf{\Theta}(z^{-1}) = \begin{bmatrix} \mathbf{\Theta}_1(z^{-1}) \\ \vdots \\ \mathbf{\Theta}_{N_p}(z^{-1}) \end{bmatrix} = \begin{bmatrix} \mathbf{\Theta}_{1,0} + \mathbf{\Theta}_{1,1} z^{-1} + \cdots + \mathbf{\Theta}_{1,nd-1} z^{-(nd-1)} \\ \vdots \\ \mathbf{\Theta}_{N_p,0} + \mathbf{\Theta}_{N_p,1} z^{-1} + \cdots + \mathbf{\Theta}_{N_p,nd-1} z^{-(nd-1)} \end{bmatrix}$$

$$\mathbf{F\Gamma\Theta} = \begin{bmatrix} \mathbf{F}_{1,0} & \mathbf{F}_{1,1} & \cdots & \mathbf{F}_{1,na} \\ \vdots & \vdots & \vdots & \vdots \\ \mathbf{F}_{N_p,0} & \mathbf{F}_{N_p,1} & \cdots & \mathbf{F}_{N_p,na} \\ \mathbf{\Gamma}_{1,0} & \mathbf{\Gamma}_{1,1} & \cdots & \mathbf{\Gamma}_{1,nb-1} \\ \vdots & \vdots & \vdots & \vdots \\ \mathbf{\Gamma}_{N_p,0} & \mathbf{\Gamma}_{N_p,1} & \cdots & \mathbf{\Gamma}_{N_p,nb-1} \\ \mathbf{\Theta}_{1,0} & \mathbf{\Theta}_{1,1} & \cdots & \mathbf{\Theta}_{1,nd-1} \\ \vdots & \vdots & \vdots & \vdots \\ \mathbf{\Theta}_{N_p,0} & \mathbf{\Theta}_{N_p,1} & \cdots & \mathbf{\Theta}_{N_p,nd-1} \end{bmatrix}$$

$$\mathbf{YUV}^f = \begin{bmatrix} \mathbf{y}^f(t) \\ \vdots \\ \mathbf{y}^f(t-na) \\ \Delta\mathbf{u}^f(t-1) \\ \vdots \\ \Delta\mathbf{u}^f(t-nb) \\ \Delta\mathbf{v}^f(t) \\ \vdots \\ \Delta\mathbf{v}^f(t-nd+1) \end{bmatrix}$$

If future values of the disturbances $\mathbf{v}(t)$ are known, now the combination $\mathbf{F}'' = \mathbf{H}'\tilde{\mathbf{V}} + \mathbf{F}'$ can be made according to the approach described on page 108 in order to further simplify the calculation of the GPC control. In this case, the modified free response results into:

$$\mathbf{F}'' = \mathbf{H}' \cdot \tilde{\mathbf{V}} + \mathbf{F\Gamma\Theta} \cdot \mathbf{YUV}^f = \mathbf{H}'\mathbf{F\Gamma\Theta} \cdot \tilde{\mathbf{V}}\mathbf{YUV}^f$$

with

$$\mathbf{H}'\mathbf{F}\mathbf{\Gamma}\mathbf{\Theta} = \begin{bmatrix} \mathbf{H}'_0 & 0 & \cdots & 0 \\ \mathbf{H}'_1 & \mathbf{H}'_0 & \cdots & 0 \\ \vdots & \vdots & \ddots & \vdots \\ \mathbf{H}'_{N_p-1} & \mathbf{H}'_{N_p-2} & \cdots & \mathbf{H}'_0 \\ \mathbf{F}_{1,0} & \mathbf{F}_{1,1} & \cdots & \mathbf{F}_{1,na} \\ \mathbf{F}_{2,0} & \mathbf{F}_{2,1} & \cdots & \mathbf{F}_{2,na} \\ \vdots & \vdots & \vdots & \vdots \\ \mathbf{F}_{N_p,0} & \mathbf{F}_{N_p,1} & \cdots & \mathbf{F}_{N_p,na} \\ \mathbf{\Gamma}_{1,0} & \mathbf{\Gamma}_{1,1} & \cdots & \mathbf{\Gamma}_{1,nb-1} \\ \mathbf{\Gamma}_{2,0} & \mathbf{\Gamma}_{2,1} & \cdots & \mathbf{\Gamma}_{2,nb-1} \\ \vdots & \vdots & \vdots & \vdots \\ \mathbf{\Gamma}_{N_p,0} & \mathbf{\Gamma}_{N_p,1} & \cdots & \mathbf{\Gamma}_{N_p,nb-1} \\ \mathbf{\Theta}_{1,0} & \mathbf{\Theta}_{1,1} & \cdots & \mathbf{\Theta}_{1,nd-1} \\ \mathbf{\Theta}_{2,0} & \mathbf{\Theta}_{2,1} & \cdots & \mathbf{\Theta}_{2,nd-1} \\ \vdots & \vdots & \vdots & \vdots \\ \mathbf{\Theta}_{N_p,0} & \mathbf{\Theta}_{N_p,1} & \cdots & \mathbf{\Theta}_{N_p,nd-1} \end{bmatrix}$$

$$\tilde{\mathbf{V}}\mathbf{Y}\mathbf{U}\mathbf{V}^f = \begin{bmatrix} \Delta\mathbf{v}(t+1) \\ \vdots \\ \Delta\mathbf{v}(t+N_p) \\ \mathbf{y}^f(t) \\ \vdots \\ \mathbf{y}^f(t-na) \\ \Delta\mathbf{u}^f(t-1) \\ \vdots \\ \Delta\mathbf{u}^f(t-nb) \\ \Delta\mathbf{v}^f(t-1) \\ \vdots \\ \Delta\mathbf{v}^f(t-nd+1) \end{bmatrix}$$

If the current values of the disturbances $\mathbf{v}(t)$ are known, but not their future trend, then they are assumed to be constant. This leads to the fact that the expression $\Delta\mathbf{v}(t)$ becomes zero and that the term $\mathbf{H}'\tilde{\mathbf{V}}$ disappears. Then, the calculation of a modified free response $\mathbf{F}''$ is unnecessary and in equation (8.50), $\mathbf{F}''$ is replaced with $\mathbf{F}'$.

8.3.4 Consideration of the control horizon

According to chapters 8.1.4 and 8.2.4 a simplification of the calculations under consideration of the control horizon N_u, $N_u \leqq N_2$ is possible, even if the filter characteristics of GPC are used. Then equation (8.53) becomes

$$\begin{bmatrix} \hat{\mathbf{y}}(t+N_1) \\ \vdots \\ \hat{\mathbf{y}}(t+N_2) \end{bmatrix} = \begin{bmatrix} \mathbf{G}'_{N_1-1} & \mathbf{G}'_{N_1-2} & \cdots & \mathbf{G}'_{N_1-N_u} \\ \vdots & \vdots & & \vdots \\ \mathbf{G}'_{N_2-1} & \mathbf{G}'_{N_2-2} & \cdots & \mathbf{G}'_{N_2-N_u} \end{bmatrix} \begin{bmatrix} \Delta\mathbf{u}(t) \\ \vdots \\ \Delta\mathbf{u}(t+N_u-1) \end{bmatrix}$$
$$+ \begin{bmatrix} \mathbf{H}'_{N_1-1} & \mathbf{H}'_{N_1-2} & \cdots & \mathbf{H}'_{N_1-N_u} \\ \vdots & \vdots & & \vdots \\ \mathbf{H}'_{N_2-1} & \mathbf{H}'_{N_2-2} & \cdots & \mathbf{H}'_{N_2-N_u} \end{bmatrix} \begin{bmatrix} \Delta\mathbf{v}(t+1) \\ \vdots \\ \Delta\mathbf{v}(t+N_u) \end{bmatrix}$$
$$+ \begin{bmatrix} \mathbf{f}'(t+N_1) \\ \vdots \\ \mathbf{f}'(t+N_2) \end{bmatrix}$$

or

$$\mathbf{Y}_{N_{12}} = \mathbf{G}'_{N_{12u}} \tilde{\mathbf{U}}_{N_u} + \mathbf{H}'_{N_{12u}} \tilde{\mathbf{V}} + \mathbf{F}'_{N_{12}} \tag{8.54}$$

The same simplifications can be carried out for equation (8.50) with which the calculation of optimum values for the actuating variables is done:

$$\tilde{\mathbf{U}}_{N_u} = ({\mathbf{G}'_{N_{12u}}}^T \mathbf{G}'_{N_{12u}} + \lambda\mathbf{I})^{-1} \cdot {\mathbf{G}'_{N_{12u}}}^T (\mathbf{W}_{N_{12}} - \mathbf{F}''_{N_{12}}) \tag{8.55}$$

Based on the fact that in most cases future values of the disturbances will be unknown, $\mathbf{F}''_{N_{12}}$ can be replaced with $\mathbf{F}'_{N_{12}}$ as already described above. Then, the matrix $\mathbf{H}'_{N_{12u}}$ is redundant, too, and it is not necessary to calculate it. Therefore, the free response $\mathbf{F}'_{N_{12}}$ is given by:

$$\mathbf{F}'_{N_{12}} = \mathbf{F}\mathbf{\Gamma}\mathbf{\Theta}_{N_{12}} \cdot \mathbf{YUV}^f$$

in which

$$\mathbf{F\Gamma\Theta}_{N_{12}} = \begin{bmatrix} \mathbf{F}_{N_1,0} & \mathbf{F}_{N_1,1} & \cdots & \mathbf{F}_{N_1,na} \\ \vdots & \vdots & \vdots & \vdots \\ \mathbf{F}_{N_2,0} & \mathbf{F}_{N_2,1} & \cdots & \mathbf{F}_{N_2,na} \\ \mathbf{\Gamma}_{N_1,0} & \mathbf{\Gamma}_{N_1,1} & \cdots & \mathbf{\Gamma}_{N_1,nb-1} \\ \vdots & \vdots & \vdots & \vdots \\ \mathbf{\Gamma}_{N_2,0} & \mathbf{\Gamma}_{N_2,1} & \cdots & \mathbf{\Gamma}_{N_2,nb-1} \\ \mathbf{\Theta}_{N_1,0} & \mathbf{\Theta}_{N_1,1} & \cdots & \mathbf{\Theta}_{N_1,nd-1} \\ \vdots & \vdots & \vdots & \vdots \\ \mathbf{\Theta}_{N_2,0} & \mathbf{\Theta}_{N_2,1} & \cdots & \mathbf{\Theta}_{N_2,nd-1} \end{bmatrix}$$

8.4 Experimental results

8.4.1 Current control

The effectiveness of considering disturbances using a multidimensional GPC controller was experimentally tested. For these tests, different reference value steps were applied to the stator current components i_{sd} and i_{sq} and the controller behavior for the other current component was observed. As figure 2.3 on page 10 shows, the cross coupling between i_{sd} and i_{sq} is proportional to the rotating frequency ω_s. Therefore, considering a machine in standstill operation ($\omega = 0$), a feedback of the rotor flux $\boldsymbol{\psi}_r$ on the stator current $\boldsymbol{i}_s$ is not existant. Because of this, the reference value for the mechanical rotating speed ω was set to $\omega = 0.5$ (half nominal speed) for the experiments; the electrical rotating speed ω_s will be slightly higher because of the slip. The used prediction horizons are $N_p = 4$ and $N_u = 2$; the degrees of the individual system polynomials have the values $na = 1$, $nb = 1$, $nd = 1$ and $nt = 1$. In the following, the system matrices for the MIMO-GPC controller are given:

$$\mathbf{G_d}(z^{-1}) = \begin{bmatrix} \frac{0{,}1484\,z^{-1}}{1-0{,}9873\,z^{-1}} & 0 & 0 \\ 0 & \frac{0{,}1484\,z^{-1}}{1-0{,}9873\,z^{-1}} & 0 \\ 0 & 0 & 0 \end{bmatrix}$$

$$\mathbf{H_d}(z^{-1}) = \begin{bmatrix} 0 & \frac{0{,}03197\,z^{-1}}{1-0{,}9873\,z^{-1}} & 0 \\ \frac{-0{,}03197\,z^{-1}}{1-0{,}9873\,z^{-1}} & 0 & \frac{-0{,}1423\,z^{-1}}{1-0{,}9873\,z^{-1}} \\ 0 & 0 & 0 \end{bmatrix}$$

$$\mathbf{T}(z^{-1}) = \begin{bmatrix} 1-0,95\,z^{-1} & 0 & 0 \\ 0 & 1-0,95\,z^{-1} & 0 \\ 0 & 0 & 1 \end{bmatrix}$$

$$\mathbf{A}(z^{-1}) = \begin{bmatrix} 1-0,9873\,z^{-1} & 0 & 0 \\ 0 & 1-0,9873\,z^{-1} & 0 \\ 0 & 0 & 1 \end{bmatrix}$$

$$\mathbf{B}(z^{-1}) = \begin{bmatrix} 0,1484\,z^{-1} & 0 & 0 \\ 0 & 0,1484\,z^{-1} & 0 \\ 0 & 0 & 0 \end{bmatrix}$$

$$\mathbf{D}(z^{-1}) = \begin{bmatrix} 0 & 0,03197\,z^{-1} & 0 \\ -0,03197\,z^{-1} & 0 & -0,1423\,z^{-1} \\ 0 & 0 & 0 \end{bmatrix}$$

$$\lambda = \begin{bmatrix} 0,003 & 0 & 0 \\ 0 & 0,003 & 0 \\ 0 & 0 & 0,003 \end{bmatrix}$$

The measurement results can be found in figure 8.1. The plots show the effect of a step in the torque-producing current component from $i_{sq} = 0$ to $i_{sq} = 1$ to the field-producing current component i_{sd}, which is controlled with the same MIMO-GPC controller. The field-producing current component is controlled to the constant value $i_{sd} = 0.2$. Figure 8.1(a) shows that during the increase of the torque-producing current component i_{sq} the flux-producing current component i_{sd} shows a recognizable deviation from its reference value when a GPC controller without consideration of disturbances is used. If the known influence of the cross coupling is considered in the controller, the control error of i_{sd} disappears nearly completely during the increase of i_{sq} (figure 8.1(b)).

8.4.2 Computation times

The computation time necessary for the multidimensional current controller with consideration of disturbances is around 4.1 µs, i. e. approximately in the same range as the calculation time for two unidimensional GPC current controllers (see chapter 6.3.4 on page 66). Since, in contrast to the two unidimensional currents controllers, the MIMO control considers the cross coupling between i_{sd} und i_{sq} as a disturbance input, multivariable control for an individual control of the flux- and torque-producing current components is clearly preferable.

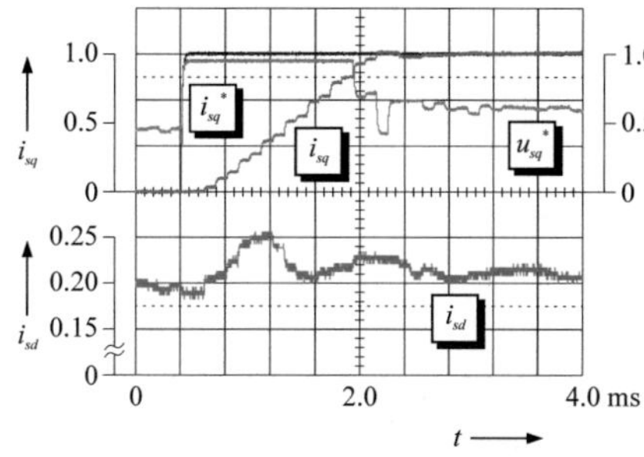

(a) Without consideration of disturbances

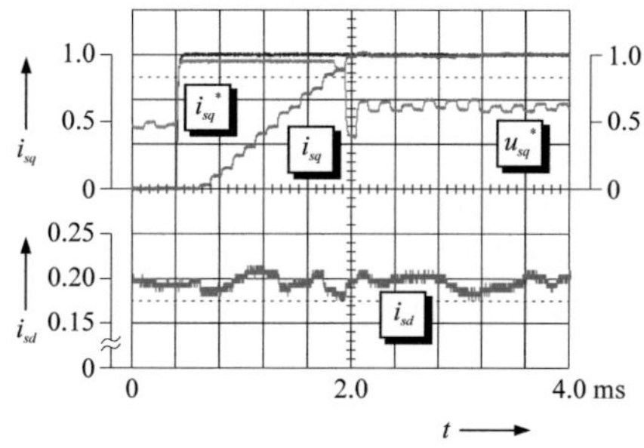

(b) With consideration of disturbances

Figure 8.1: Multidimensional current control

8.5 Comparative summary

A mathematical view of the different variations of MIMO-GPC shows the exact differences and is helpful for the interpretation of the results shown above. Table 8.1 gives an overview of the various equations for the calculation of the free response and of future optimum values for the actuating variables in different configurations.

	without disturbances	with disturbances
without filter	$\mathbf{F} = \mathbf{F}(z^{-1})\mathbf{y}(t)$ $+ \mathbf{G_p}(z^{-1})\Delta\mathbf{u}(t-1)$ $= \mathbf{FG_p} \cdot \mathbf{YU}$ $\tilde{\mathbf{U}} = (\mathbf{G}^T\mathbf{G} + \lambda\mathbf{I})^{-1}$ $\cdot \mathbf{G}^T(\mathbf{W} - \mathbf{F})$	$\mathbf{F} = \mathbf{F}(z^{-1})\mathbf{y}(t)$ $+ \mathbf{G_p}(z^{-1})\Delta\mathbf{u}(t-1)$ $+ \mathbf{H_p}(z^{-1})\Delta\mathbf{v}(t)$ $= \mathbf{FG_pH_p} \cdot \mathbf{YUV}$ $\mathbf{F}' = \mathbf{H} \cdot \tilde{\mathbf{V}} + \mathbf{F}$ $\tilde{\mathbf{U}} = (\mathbf{G}^T\mathbf{G} + \lambda\mathbf{I})^{-1}$ $\cdot \mathbf{G}^T(\mathbf{W} - \mathbf{F}')$
with filter	$\mathbf{F}' = \mathbf{F}(z^{-1})\mathbf{y}^f(t)$ $+ \mathbf{\Gamma}(z^{-1})\Delta\mathbf{u}^f(t-1)$ $= \mathbf{F\Gamma} \cdot \mathbf{YU}^f$ $\tilde{\mathbf{U}} = (\mathbf{G}'^T\mathbf{G}' + \lambda\mathbf{I})^{-1}$ $\cdot \mathbf{G}'^T(\mathbf{W} - \mathbf{F}')$	$\mathbf{F}' = \mathbf{F}(z^{-1})\mathbf{y}^f(t)$ $+ \mathbf{\Gamma}(z^{-1})\Delta\mathbf{u}^f(t-1)$ $+ \mathbf{\Theta}(z^{-1})\Delta\mathbf{v}^f(t)$ $= \mathbf{F\Gamma\Theta} \cdot \mathbf{YUV}^f$ $\mathbf{F}'' = \mathbf{H}' \cdot \tilde{\mathbf{V}} + \mathbf{F}'$ $\tilde{\mathbf{U}} = (\mathbf{G}'^T\mathbf{G}' + \lambda\mathbf{I})^{-1}$ $\cdot \mathbf{G}'^T(\mathbf{W} - \mathbf{F}'')$

Table 8.1: Comparison of different MIMO-GPC variations

As already described in chapter 8.2.3 on page 99 et seqq., $\mathbf{G}$ and $\mathbf{G}'$ are absolutely identical, no matter if disturbance inputs are taken into consideration, or not. Thus, the only difference between the structures of MIMO-GPC controllers with and without consideration of known disturbances is the calculation of the free response $\mathbf{F}$ or $\mathbf{F}'$. Thereby only the term $\mathbf{H_p}(z^{-1})\Delta\mathbf{v}(t)$ or $\mathbf{\Theta}(z^{-1})\Delta\mathbf{v}^f(t)$ is added since $\mathbf{F}$, $\mathbf{G_p}$, $\mathbf{F}'$ and $\mathbf{\Gamma}$ are all independent of the consideration of disturbances. Examining the values in the polynomial matrix $\mathbf{\Theta}(z^{-1})$, it can be seen that the influence of the disturbances on the calculation of the free response is quite small which, however, describes the actual behavior of the machine. Thus, the behavior of the controller does not change much if the disturbances are taken into account in the GPC controller. Despite this, as shown in figure 8.1, a significant suppression of the cross coupling is noticeable.

9 Direct model-based predictive control

All the techniques for optimum control of a drive by means of Long-Range Predictive Control (LRPC), which have been described so far, are based on the principle that an optimum voltage space vector is determined, which will be applied to the machine being controlled in the next switching cycle. Commonly a two-level inverter fed from a DC link voltage is used as voltage source for the machine, as shown in figure 9.1. This inverter consists of three half bridges, labeled with the letters a, b and c. They can independently connect the three phases of the machine either to the positive or to the negative DC link voltage u_d. Hence, it follows that only a finite number of possible voltage space vectors $\boldsymbol{u}_s$ can exist. In control engineering, such systems, consisting of interdependent physical laws, logical rules and which are subject to constraints of the operating range, i. e. their input variables, state variables and output variables can assume partly continuous but also partly discrete values, are named *Mixed Logical Dynamical Systems (MLD)* or *Hybrid Systems*.

In drive control, the normal approach is that the controller output is fed to a so-called modulator which discretizes the reference space voltage vector (figure 9.2(a)). This means that the modulator synthesizes the desired space voltage vector which can normally not be delivered from the inverter out of three other space voltage vectors that can be delivered. The value of these voltage space vectors averaged over one sampling interval is equal to the desired reference value. As conventional model-based techniques do not consider this fact, the actual optimum control is impaired by the modulator which is neither considered in the controller design nor in the model. An optimum control which regards the fact that the inverter can realize only a finite number of possible output states, must be superior to the MPC techniques discussed so far. Hence, a new MPC scheme is introduced in the following: It allows direct inverter control, i. e. the control algorithm directly determines the optimum inverter switching state for the next sampling cycle. This technique is called *Direct Model Predictive Control* or *DMPC* (figure 9.2(b)).

The basic functional principle of direct model-based predictive control can be explained best with an example of playing chess. The chess player precalculates all discrete values of the variables, in this case the moves of his chessmen, up to

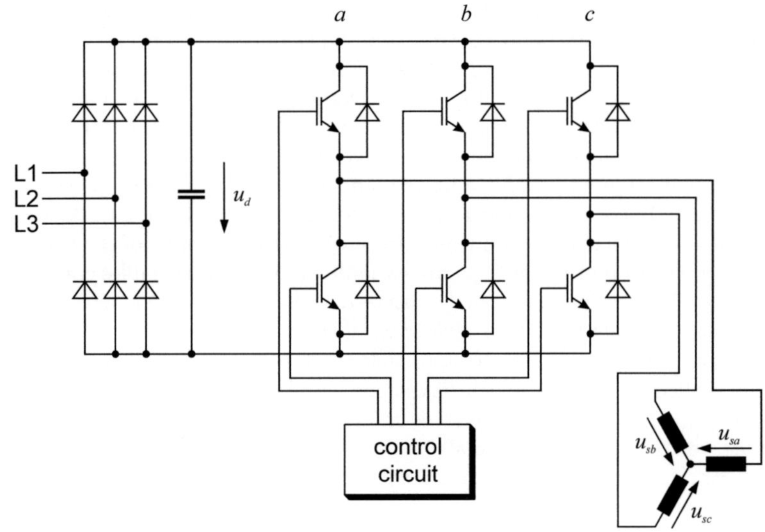

Figure 9.1: Two-level inverter circuit with DC link

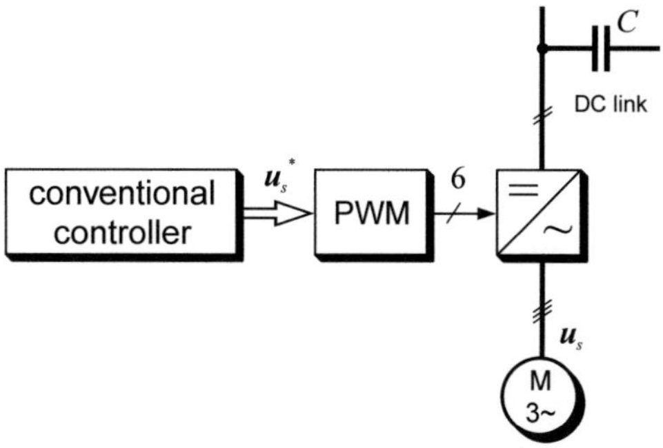

(a) conventional

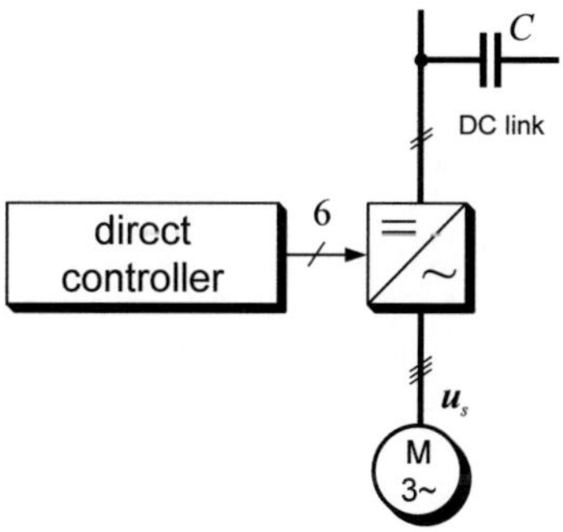

(b) direct

Figure 9.2: Principle of a machine control

a "prediction horizon" for the further match, including the possible movements of his opponent, i. e. the behavior of the controlled system. Besides, he takes his decisions similar to the Branch and Bound principle (see chapter 9.4.1) because movement combinations which lead recognizably to a disadvantage for him are not examined further, instead he immediately looks for alternatives. After finishing the optimization, he will choose the best possible move for him, carry it out and then wait for the reaction of the opponent which corresponds again to the principle of Receding Horizon. After the system reaction, i. e. the opponent's move, the whole precalculation and optimization will be started again.

9.1 Published techniques

As already mentioned in chapter 5, model-based control schemes are already known for more than 30 years in control engineering and are also applied in practical applications. A short historical overview is given e. g. in the publications of Clarke [30] and Morari/Lee [88]. Nevertheless, the implicit consideration of switching actuators has barely found an entree in these published control techniques. The work carried out at the Technical University of Aachen in the 1980s by Hoffmann [51, 52] and Schmitz [105] can be taken as an exception. Just recently, research in the control of hybrid systems has gained in importance and now publications which deal with the application of model-based predictive techniques to these systems are known [10].

In 1984, Hoffmann published in his dissertation [51] under the name *Adaptive Two-step Control* for the first time an MPC control scheme which considered the special constraints of an actuator with a finite number of discrete switching states already in the approach itself. Thereby, Hoffmann only dealt with a two-step controller. The description of the control strategy shows clearly that it is a typical MPC control scheme:

- The controller itself contains a system model. The model parameters can be adapted online.
- The future behavior of the system is precalculated for all possible sequences of control values using a predictor and it is evaluated via a cost function. The sequence of control values with the best result is selected and the *first* value of this sequence is applied to the controlled system.
- The advance calculation will be done up to a predetermined cost and control horizon.

As the above points prove, Adaptive Two-step Control is definitely a special case of Direct Model Predictive Control.

The two-step control by Hoffmann was further developed by Schmitz and published as *Adaptive Predictive Three-step Control* [105]. Besides the extension of the method to a three-step control, Schmitz has carried out additional improvements in detail, among other things, with regard to the search for an optimum sequence of control values. Special attention was given to the reduction of the necessary computation time, an aspect which can, even today, be a decisive factor for the feasibility of a DMPC control.

9.2 Inverter operation with DMPC

In drive technology, in most cases, a two-level inverter fed from a DC link is used as an actuator for closed-loop current control. As it is commonly known, such an inverter has eight discrete switching states which can directly generate seven different voltage vectors. The possible voltage values can be calculated with the equations (2.1) to (2.4) given in chapter 2 on page 7. Figure 9.3 shows the results in space vector representation; the corresponding values are depicted in table 9.1. Of course, the DMPC method can be applied to any kind of inverter with a higher number of switching levels and of voltage or current source type. Then, however, figure 9.3 and table 9.1 must be modified accordingly. It has to be pointed out that the number of possible voltage space vectors and thus, the number of possible switching sequences to be precalculated will rise exponentially when multilevel inverters are considered.

When controlling an electrical machine, the controller can be implemented either in stationary or in rotating coordinates. Since the inverter switching states seen from a rotating coordinate frame will be constantly changing, it is better to realize a direct controller in a stationary (stator) coordinate system. Thus the reference values will change sinusoidally with time, however, with an MPC controller this does not lead to a contouring error as long as the sine form of the reference value can be precalculated and then be forwarded to the controller input. Although some additional computation time is necessary, this is less time consuming than the transformation of all the possible switching states of the inverter into the rotating coordinate frame and their following forecast. Furthermore, predicting the sine form of the reference current value can simply be omitted at the expense of a contouring error, i. e. for the complete prediction horizon N_p, it can be assumed that the reference value is constant. Implementing the DMPC controller in field coordinates on the other hand,

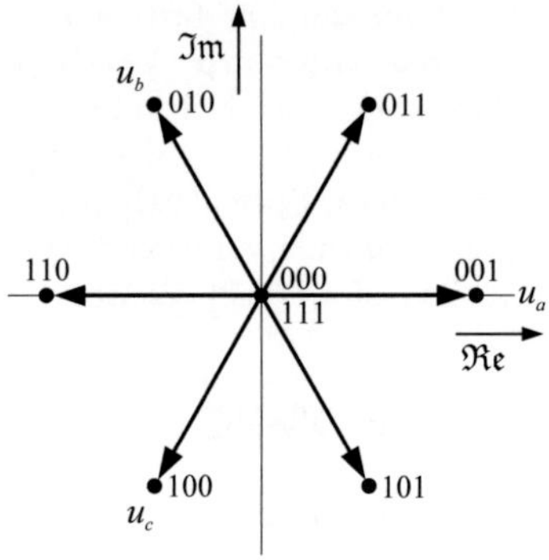

Figure 9.3: Possible switching states of a two-level inverter

Vector number	Switching state Bridge c	Bridge b	Bridge a	Voltage $u_{s\alpha}$	$u_{s\beta}$
0	0	0	0	0	0
1	0	0	1	$\frac{2}{\sqrt{3}}$	0
2	0	1	0	$-\frac{1}{\sqrt{3}}$	1
3	0	1	1	$\frac{1}{\sqrt{3}}$	1
4	1	0	0	$-\frac{1}{\sqrt{3}}$	-1
5	1	0	1	$\frac{1}{\sqrt{3}}$	-1
6	1	1	0	$-\frac{2}{\sqrt{3}}$	0
7	1	1	1	0	0

Table 9.1: Possible switching states of a two-level inverter

would mean that a prediction of the future location of the possible inverter switching states would be *absolutely necessary* since otherwise no control would be possible.

9.2.1 Consideration of the Bootstrap circuit

Low-cost inverters often have a so-called *bootstrap circuit* as power supply for the driver components of the IGBT module. When such a circuit is used, a power supply for the driver circuits of the "upper" half-bridge IGBT switches with special DC/DC converters is not required; unfortunately, this leads to the serious drawback that the "upper" IGBT drivers are only supplied with power if the "lower" half-bridge IGBT switches are turned on. This fact can be handled when the corresponding driver modules are equipped with a storage capacitor which backs up the power supply while the "upper" IGBTs are turned on. Since this storage capacitor is only recharged when the "lower" devices are switched on, the duty cycles cannot be arbitrarily chosen. By using a modulation technique, e. g. space vector modulation, it can be ensured that the "lower" semiconductor devices are switched on often enough to guarantee power supply for the driver of the "upper" switches by limiting the modulation index to e. g. $m_{max} = 0.95$. A direct inverter control method without the need for a modulator does not have this possibility. Nevertheless, in this case it also has to be ensured that after a certain on-time of the "upper" device, the "lower" device will be switched on for at least one sampling cycle in order to recharge the capacitor for the power supply of the "upper" IGBT drivers.

Classical direct control methods, e. g. Direct Torque Control (DTC) [1, 113] or Direct Self Control (DSC) [33], offer no such possibility to pass the information to the controller which voltage vectors should not be switched due to the aforementioned constraints of the bootstrap circuit. In these cases, the easiest method is simply to switch on the "lower" device of the related half bridge compulsorily and thus, to ignore the optimum voltage vector precalculated by the control algorithm at least partially. However, this leads to a non-optimum control since by the *subsequent* correction of the optimum vector it cannot be guaranteed that the best possible switching state has been selected from the remaining allowed ones.

Model-based control strategies (MPC controllers) are predestined for applications in which constraints of the actuator or of the plant have to be considered already in the controller design. Accordingly, when using DMPC methods, it is possible to pass the information about the constraint resulting from the bootstrap circuit to the controller already *beforehand*. Terno [115, p. 28] defines this as a *prohibited family*, i. e. a subset of possible solutions of the optimization problem which may not be used because of any constraints and hence, are not available as an optimum solution. The knowledge at which time some voltage vectors may not be switched can be included in the design of the control algorithm and hence, the controller *will consider this for the whole control horizon in advance*. Because of this, it is ensured that the controller will find the optimum switching sequence among the remaining possibilities under all conditions. Therefore DMPC techniques are superior to *any* other direct control method.

9.3 Model formulation

Because of the close relationship with conventional MPC, modeling can be done following the same principles. In [51] Hoffmann also uses an approach similar to a CARIMA model. However, considering the derivation of the machine model one has to take care that a DMPC controller—as mentioned in chapter 9.2—is better realized in a stationary (stator) coordinate system.

Basic for all models used to control an induction machine is always the complex machine model as stated in chapter 2, which can be described by equations (2.9) and (2.10) (see also figure 2.2 on page 9). If a stationary coordinate frame is selected, i. e. $\omega_k = 0$, and the equations (2.9) and (2.10) are splitted into real and imaginary part, the following system of equations can be obtained:

$$i_{s\alpha} + \tau_\sigma{}' \frac{di_{s\alpha}}{d\tau} = \frac{k_r}{r_\sigma \tau_r}\psi_{r\alpha} + \frac{k_r}{r_\sigma}\omega\psi_{r\beta} + \frac{1}{r_\sigma}u_{s\alpha} \tag{9.1}$$

$$i_{s\beta} + \tau_\sigma{}' \frac{di_{s\beta}}{d\tau} = \frac{k_r}{r_\sigma \tau_r}\psi_{r_\beta} - \frac{k_r}{r_\sigma}\omega\psi_{r\alpha} + \frac{1}{r_\sigma}u_{s\beta} \tag{9.2}$$

$$\psi_{r\alpha} + \tau_r \frac{d\psi_{r\alpha}}{d\tau} = -\omega\tau_r\psi_{r\beta} + l_h i_{s\alpha} \tag{9.3}$$

$$\psi_{r\beta} + \tau_r \frac{d\psi_{r\beta}}{d\tau} = \omega\tau_r\psi_{r\alpha} + l_h i_{s\beta} \tag{9.4}$$

In order to clarify the relations between $i_{s\alpha}$, $i_{s\beta}$, $\psi_{r\alpha}$ and $\psi_{r\beta}$, figure 9.4 shows the signal flow graph of an induction machine in stationary coordinates.

In the same way as with continuous signal MPC controllers, the control algorithm can be developed as a single or MIMO controller; consideration of disturbances and filtering of noisy measurement signals can be implemented, too. Since, in these cases, the approach is identical to the ones described in chapters 6 and 8, only one of the many possible combinations is discussed here:

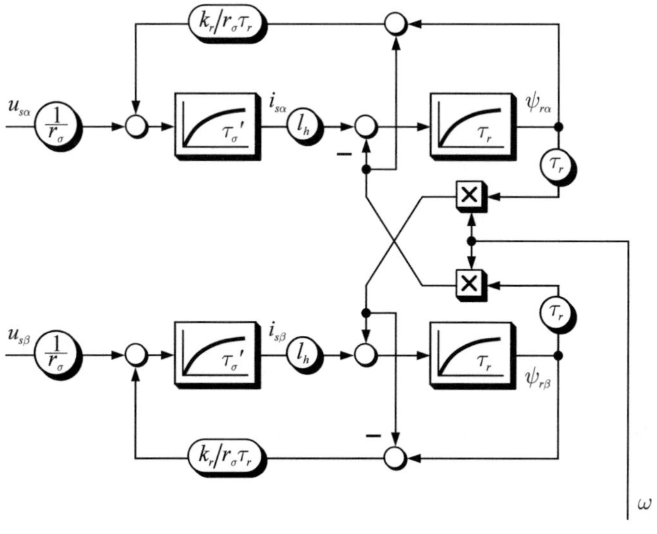

Figure 9.4: Scalar machine model (stator coordinates)

A MIMO-DMPC algorithm is introduced in which the cross coupling between $\psi_{r\alpha}$ and $\psi_{r\beta}$ as well as the impact of $i_{s\alpha}$ and $i_{s\beta}$ caused by the rotor flux (figure 9.4) is neglected due to simplicity.

As the signal flow graph in figure 9.4 shows, the stator voltages $u_{s\alpha}$ and $u_{s\beta}$ are the input variables of the plant and thus, they are also the actuating variables. A MIMO-DMPC controller for current control of an induction machine with a DC link inverter, using this model with $u_{s\alpha}$ and $u_{s\beta}$ as plant inputs, has to optimize two actuating variables, which can, according to table 9.1, take only a few discrete values. In addition to that, $u_{s\alpha}$ and $u_{s\beta}$ are not independent from each other because they depend on the switching states of the inverter half bridges. Furthermore, the change of the actuating values $\Delta\mathbf{u}$ does in this case *not* express the real effort for changing the actuating variables because the effort is not dependent on the change of the voltage level of $u_{s\alpha}$ or $u_{s\beta}$, but only on the switching operations in the inverter itself. Because of this, it is reasonable to use the switching states of the three half bridges as actuating variables instead of the complex stator voltages. As a matter of course, in this case, the conversion from the half bridge states to the machine voltages $u_{s\alpha}$ and $u_{s\beta}$ has to be added to the plant model. The necessary conversion factors can be taken from table 9.1. Such an approach has several advantages:

- The cost function evaluates the real switching effort, i. e. the switchings that have to be executed in the inverter. This means that only by selecting the appropriate weighting factor, the average switching frequency of the DMPC controlled inverter can be lowered or raised. The controller then tries to achieve an optimal result for the whole cost function, consisting of control deviation and switching operations, by switching other bridges.

- The actuating variables directly correspond to the states of the single half bridges. Hence, the inverter states are directly optimized. The results can be forwarded to the inverter without any further conversion; an additional mathematical transformation is not necessary.

- A two-level inverter with DC link permits eight different switching states. These can, however, generate only seven different space voltage vectors. Choosing the inverter states as actuating variables, it is possible to consider all eight switching possibilities of the inverter. Dependent on the better alternative concerning the switching frequency, the zero voltage space vector is created either by the switching state 000 or 111. On the contrary, if the stator voltages $u_{s\alpha}$ and $u_{s\beta}$ are selected as actuating variables, this degree of freedom is lost.

- Considering an inverter with DC link, every half bridge has only two possible states. Thus, each of the three actuating variables can take only two different states. Hence, the optimization problem is a so-called boolean optimization. For this kind of optimization, there exists a wide range of simplifications which lead to efficient optimization algorithms.

9.3.1 Simple machine model

A simple machine model without consideration of the cross coupling or of disturbances makes it easier to get familiar with direct model-based predictive control. Thus, the machine model is reduced to two single and independent first order transfer functions as it can be seen in figure 9.5.

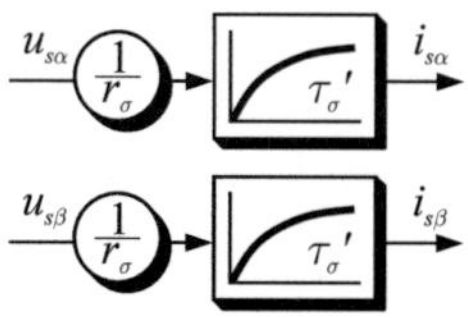

Figure 9.5: Simplified model for current control (stator coordinates)

Since, as mentioned above, not the stator voltages $u_{s\alpha}$ and $u_{s\beta}$, but the switching states a, b and c of the individual half bridges should be selected as input variables, the corresponding translation has to be added to the machine model shown above (figure 9.6). The translation block from a, b, c to $u_{s\alpha}, u_{s\beta}$ contains the factors given in table 9.1.

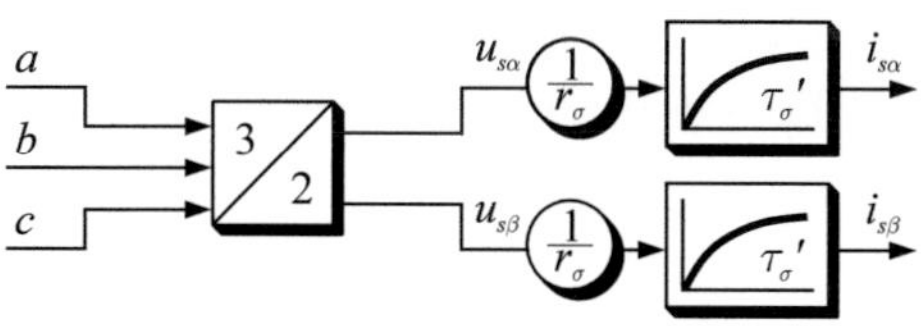

Figure 9.6: Simplified model for current control (bridge switching states)

If the signal flow graph shown in figure 9.6 is translated into state space equations according to equations (5.2) and (5.3) (see page 31), the vectors $\mathbf{x}$,

u and **y** are selected as follows:

$$\mathbf{x} = \begin{bmatrix} i_{s\alpha} \\ i_{s\beta} \end{bmatrix}, \qquad \mathbf{u} = \begin{bmatrix} a \\ b \\ c \end{bmatrix}, \qquad \mathbf{y} = \begin{bmatrix} i_{s\alpha} \\ i_{s\beta} \end{bmatrix}$$

Thus, the state space representation of figure 9.6 is given as:

$$\begin{aligned} \frac{d}{d\tau}\begin{bmatrix} i_{s\alpha} \\ i_{s\beta} \end{bmatrix} &= \begin{bmatrix} -\frac{1}{\tau_\sigma{}'} & 0 \\ 0 & -\frac{1}{\tau_\sigma{}'} \end{bmatrix} \cdot \begin{bmatrix} i_{s\alpha} \\ i_{s\beta} \end{bmatrix} \\ &+ \begin{bmatrix} \frac{2}{\sqrt{3}r_\sigma\tau_\sigma{}'} & -\frac{1}{\sqrt{3}r_\sigma\tau_\sigma{}'} & -\frac{1}{\sqrt{3}r_\sigma\tau_\sigma{}'} \\ 0 & \frac{1}{r_\sigma\tau_\sigma{}'} & -\frac{1}{r_\sigma\tau_\sigma{}'} \end{bmatrix} \cdot \begin{bmatrix} a \\ b \\ c \end{bmatrix} \end{aligned} \tag{9.5}$$

$$\begin{bmatrix} i_{s\alpha} \\ i_{s\beta} \end{bmatrix} = \begin{bmatrix} 1 & 0 \\ 0 & 1 \end{bmatrix} \cdot \begin{bmatrix} i_{s\alpha} \\ i_{s\beta} \end{bmatrix} + \begin{bmatrix} 0 & 0 & 0 \\ 0 & 0 & 0 \end{bmatrix} \cdot \begin{bmatrix} a \\ b \\ c \end{bmatrix} \tag{9.6}$$

9.4 Implicit solution

A simple way for designing a direct control method of an inverter without an intermediary modulator under retention of Long-Range Predictive Control is to adapt a suitable GPC controller. Multidimensional controllers with and without filtering of the measured values as well as consideration of disturbances are possible without any problems. It is advisable to filter the measured stator currents with a low-pass filter since, otherwise, the control will be very difficult to handle because of the typical characteristics of GPC methods (see chapter 6.1.2 on page 48). Thus, the state space representation given in the equations (9.5) and (9.6) can be rewritten in the following way:

$$\begin{aligned} \frac{d}{d\tau}\begin{bmatrix} i_{s\alpha} \\ i_{s\beta} \end{bmatrix} &= \begin{bmatrix} -\frac{1}{\tau_\sigma{}'} & 0 \\ 0 & -\frac{1}{\tau_\sigma{}'} \end{bmatrix} \cdot \begin{bmatrix} i_{s\alpha} \\ i_{s\beta} \end{bmatrix} \\ &+ \begin{bmatrix} \frac{2}{\sqrt{3}\,r_\sigma\tau_\sigma{}'} & -\frac{1}{\sqrt{3}\,r_\sigma\tau_\sigma{}'} & -\frac{1}{\sqrt{3}\,r_\sigma\tau_\sigma{}'} \\ 0 & \frac{1}{r_\sigma\tau_\sigma{}'} & -\frac{1}{r_\sigma\tau_\sigma{}'} \end{bmatrix} \cdot \begin{bmatrix} a \\ b \\ c \end{bmatrix} \end{aligned} \tag{9.7}$$

$$\begin{bmatrix} i_{s\alpha} \\ i_{s\beta} \\ Dummy \end{bmatrix} = \begin{bmatrix} 1 & 0 \\ 0 & 1 \\ 0 & 0 \end{bmatrix} \cdot \begin{bmatrix} i_{s\alpha} \\ i_{s\beta} \end{bmatrix} + \begin{bmatrix} 0 & 0 & 0 \\ 0 & 0 & 0 \\ 0 & 0 & 0 \end{bmatrix} \cdot \begin{bmatrix} a \\ b \\ c \end{bmatrix} \tag{9.8}$$

According to that, state, input and output vector are

$$\mathbf{x} = \begin{bmatrix} i_{s\alpha} \\ i_{s\beta} \end{bmatrix}, \qquad \mathbf{u} = \begin{bmatrix} a \\ b \\ c \end{bmatrix}, \qquad \mathbf{y} = \begin{bmatrix} i_{s\alpha} \\ i_{s\beta} \\ Dummy \end{bmatrix}$$

The "dummy" variable in the output vector $\mathbf{y}$ is necessary for the internal filtering of GPC. Since it is not reasonable to use a GPC controller without internal filtering via a $\mathbf{T}(z^{-1})$ design polynomial—as already mentioned—, at least a smooth filtering via $\mathbf{T}(z^{-1})$ has to be done. Since the actuating variables $\Delta\mathbf{u}(t)$ as well as the controlled variables $\mathbf{y}(t)$ have to be filtered with the same polynomial, both vectors need to have the same dimension. Since $\mathbf{u}(t)$ has the dimension three, a "dummy" value has to be added to $\mathbf{y}(t)$ in order to get the same dimension and to make use of an identical filter for $\mathbf{y}(t)$.

In the method presented here from the free and the forced response calculated using the j-step ahead predictor and from future values of the reference variable, an optimal sequence of values for the actuating variables is determined by using a cost function. In this case, direct control does not differ from conventional MPC. Hence, the same quadratic cost function as denoted in chapter 6.1.1 on page 42 in equation (6.6) can be used. As proven in chapter 8.1.3 on page 88 this cost function is applicable for a unidimensional as well as for a multidimensional controller (equation (8.14)). Since the direct controller, in this case, should be a MIMO controller, in the following, only the multidimensional case will be dealt with. Equations and statements valid for SISO methods can easily be derived.

Analytical minimization of the cost function J leads to equation (8.17) for optimal future values of the actuating variables. However, in this case the actuating variable $\mathbf{u}(t)$ can take any value. This equation is called *single-step optimization.* If the inverter is controlled directly without an intermediary modulator a single-step optimization is not possible because the actuating variable can take only a few discrete values. Seen from the view of mathematics, this means that the actuating variable $\mathbf{u}(t)$ has to be an element of the set of possible actuating variables $\boldsymbol{S}$, which itself is a subset of the natural numbers $\mathbb{N}$, i. e. $\mathbf{u}(t) \in \boldsymbol{S} \subseteq \mathbb{N}$. It is therefore necessary to solve the optimization problem using a *multi-step optimization.* Unfortunately, a multi-step optimization, in contrary to a single-step optimization, cannot be done analytically anymore because of the nonlinearity caused by the discrete actuating variables. Thus, a decision and search strategy which can find the best possible sequence of values for the actuating variables has to be used, i. e. the smallest value of the cost function (8.16) under consideration of the special constraints for $\tilde{\mathbf{U}}$ has to be found. Since only the value of $\tilde{\mathbf{U}}$ which leads to a minimized J has to be searched for, the constant term $(\mathbf{W}-\mathbf{F})^T(\mathbf{W}-\mathbf{F})$ in equation (8.16) can be omitted; thus, the optimization rule is:

$$J = \min_{\tilde{\mathbf{U}}} \left\{ \tilde{\mathbf{U}}^T(\mathbf{G}^T\mathbf{G} + \lambda\mathbf{I})\tilde{\mathbf{U}} + 2\tilde{\mathbf{U}}^T\mathbf{G}^T(\mathbf{F}-\mathbf{W}) \right\} \tag{9.9}$$

9.4.1 Solving algorithms

For the search for the minimum of J, several methods exist. In order to decide which discrete optimization method should be used, the demands for the optimization have to be defined. The following criterions play a role for this decision (see also [105, chapter 5.3.3] and [115, chapter 2]):

- Can finiteness of the algorithm be guaranteed?
- Which computation time is necessary for the algorithm?
- If the algorithm does not deliver a solution in a defined time, does it deliver at least approximate solutions instead?
- How much memory is needed?
- Has the algorithm a clear structure and is it easily readable (structured programming)?
- Does the algorithm find the *global* and not only a *local* minimum in each case?

In the following chapters different discrete optimization methods are described and discussed under the aspect of their use in a DMPC controller for current control of an induction machine. A good overview about the algorithms is also given in the publication by Terno [115].

Exhaustive Search (Complete enumeration)

The most simple method is based on the principle that all possible sequences of values of the actuating variables $\tilde{\mathbf{U}}$ are determined first and then, one after another is evaluated with the cost function J. The sequence which leads to the smallest value of the cost function is the solution that was searched. Since in this method, all possibilities are tested, regardless of the computational effort, this method is called *exhaustive search* or *complete enumeration* in technical literature. As it can easily be seen, this strategy will always find the global minimum because all possible combinations are tested. Using the example of a two-position actuator, a bang-bang GPC controller is described by Tsang/Clarke [119]; the authors also recommend the simple exhaustive search for solving the optimization problem. However, it is questionable if this can also be recommended for systems with multilevel actuators (e. g. three-phase inverters) and for longer control horizons, or if other optimization techniques are better suited in these cases since the computation time for an exhaustive search should not be underestimated. For an actuator with x discrete actuating possibilities and a control horizon N_u exactly x^{N_u} different possible switching sequences exist, i. e. the needed computation time increases exponentially with the number of predicted time steps. For a two-level inverter with $x = 8$ different switching states and $N_u = 2$ already 64, with $N_u = 3$ as many as 512 switching sequences have to be calculated individually. As the results shown in chapter 9.4.4 prove, an implementation of an exhaustive search strategy is not feasible in drive control.

Speed-up strategy: To improve the complete enumeration method, Hoffmann has extended it by a so-called speed-up strategy [51, chapter 3.3.1] in order to save computation time. In the dynamic case, i. e. when large steps of the reference value occur, an actuating variable which moves the controlled variable to the reference value with maximum speed is applied first. If an opposite state of the actuating variable is switched at an adequate time t_1, the reference value can be reached in a minimal amount of time without any overshoot. In the speed-up strategy by Hoffmann the point of time t_1 for switching and the whole duration of the process are calculated in advance and hence, in transient state, the necessary computation time can be reduced significantly.

If the speed-up strategy by Hoffmann is compared to the trajectory-based predictive controllers, which are known in drive control and which are presented in chapter 4.1, one can see that both methods are based on the same principles. In both cases a system trajectory is selected by choosing an appropriate switching state of the value-discrete actuator and then an optimum switching time is determined. Taking conventional trajectory-based methods into consideration, at this predetermined time, another, also predetermined switching state will be applied to the plant, whereas the speed-up strategy switches back to direct model-based predictive control with exhaustive search for the optimal switching state after two steps. However, both strategies are based on the basic principle of trajectory-based control.

For DMPC of an induction machine, the speed-up strategy is not interesting since the current controller only needs few sampling cycles for reaching the reference value. Even a step of $i_{sq} = 0$ to $i_{sq} = 1.0$ at $\omega = 0.5$ takes just 13 sampling cycles. Thus, the effort for the speed-up strategy is not worthwhile.

Taking speed control into consideration, the time necessary for reaching the reference value after a step change is indeed significantly larger, but even in the stationary case still a relatively high prediction horizon is necessary (see chapter 6.3.3 on page 63 et seqq.). Thus the implementation of a speed-up strategy for this application does not lead to a decrease of the maximum necessary online computation time, too.

Branch and Bound

In his publication [105, chapter 5.2.2], Schmitz proposes a computation time saving optimization method called *Branch and Bound.* It is based on the principle that the decision tree is run through from the root to the last node only for the first path. The value J of the cost function valid for this sequence of values of the actuating variables is saved as the temporary minimum. Now, other paths through the decision tree are only examined further if they are promising for a new minimum. This means that, if the temporary value of the cost function of a partly run through path is already larger than the current, temporary minimum, the currently investigated path including all further nodes cannot contain the global minimum. Thus, the correspondent part of the decision tree does not have to be run through anymore and will be cut out. The same happens if the part of the tree which is still to be investigated belongs to a prohibited family and thus, can also not contain the optimum possible solution [115]. If the algorithm has run through a possible path to the end node and if the value of the cost function for this sequence is smaller than the temporary minimum, the temporary minimum will be replaced by the value of the cost function for the currently investigated path of the decision tree. This search strategy also guarantees that the global minimum will be found in any case.

Especially with high control horizons N_u the Branch and Bound strategy leads to a significant reduction of the computation time because branches of the decision tree which will recognizably not lead to a minimum are not investigated. A further reduction of computation time is achieved when investigating other branches of the same subtree since the temporary costs up to the current node are already known; thus, the part of the costs created by the nodes lying between the root of the decision tree and the current node does not have to be calculated again. Furthermore, additional computation time can be saved if an estimation about the minimum reachable value of the cost function in a branch is done in advance. Is this value higher than the temporary minimum, an investigation of the branch is not necessary because it will not lead to an optimum solution.

As chapter 9.4.4 shows, an online calculated solution of the optimization problem via Branch and Bound is also too slow for use in drive control. Additionally, it is to mention that the Branch and Bound principle has the drawback that an actual reduction of the computation time *cannot be guaranteed*! In the worst case it can happen that the cost function has to be evaluated for *all possible* sequences.

Cutting Planes

The Cutting Planes stragegy can be explained with few words. The basic idea is to determine the optimum solution neglecting the fact, that the solution has to be integer. The first step is to check if the obtained solution is already an element of the valid set $\boldsymbol{S} \subseteq \mathbb{N}$. If this is not the case, constraints, which cut off parts of the area of possible solutions, are added one by one; these constraints are called *cutting planes*. Then, the new linear optimization problem is solved under consideration of the constraints. These steps are repeated until an optimum solution which fulfills the integer constraint is found. A detailed description of this method can be found in the book of Terno [115, chapter 2.4] and also with Hadley [47, from chapter 8.11 on].

However, methods based on cutting planes have the fundamental drawback that they converge extremely slow [63, chapter 1.6.4]. Thus, they cannot be used in drive control.

Expansion Strategy

Besides the methods already mentioned, Terno also refers to the so-called *Expansion Strategy* [115, chapter 2.3]. This method was developed by Schoch at the Mining University (Bergakademie) Freiberg in 1970; in [106] it is described in detail. The strategy solves the optimization task by using a substitution problem. This substitution problem is defined in a subset $x \in \boldsymbol{R} \cap \boldsymbol{S}$ of the set $x \in \boldsymbol{S}$ which is valid for the original problem. Additionally, the substitution problem should fulfill the constraints that $\boldsymbol{R} \cap \boldsymbol{S} \neq \emptyset$, that a lower boundary exists for all values of the substitution problem and that the minimum of the substitution problem defined for $x \in \boldsymbol{R} \cap \boldsymbol{S}$ represents the minimum of the original optimization task for $x \in \boldsymbol{S}$. Now subsets $\boldsymbol{U}_k \subset \boldsymbol{R}$ which are defined for the substitution problem via an upper and a lower boundary are examined. The elements of $\boldsymbol{U}_k$ are tested if they are members of $\boldsymbol{S}$. The procedure is repeated until an optimum element is found. Hence, in this method the set of the examined elements is expanded step by step which leads to the name *Expansion Strategy*.

As further mentioned by Schoch, Cutting Planes as well as Branch and Bound are special cases of the Expansion Strategy [106, chapter 1.2.2 and 1.3.2].

Discrete Dynamic Programming

For the solution of a separable optimization problem via *Dynamic Programming*, the optimization problem is separated into several sub-problems which are treated one after another. This leads to a multilevel decision process with the output state of one level being the input state of the next one. This means that a multidimensional optimization task is splitted into several unidimensional sub-problems which will be solved successively; it's solved best in a recursive matter. According to Hoffmann [51, chapter 3.3.3], Dynamic Programming can indeed save computational time, however it is less suitable because of its highly increased memory requirements. For this reason it is not discussed any further; more detailed descriptions can be found e. g. in Cooper/Cooper [31].

9.4.2 Mathematical derivation

The DMPC controller with implicit solution of the optimization task is, as already explained, a GPC strategy in which the discrete-valued structure of the actuator is considered. Since this affects only the method of optimization, the mathematical derivations presented in chapter 8 are valid without any modifications. The calculation of the system matrices of the CARIMA model, the j-step ahead predictor etc. are described there. The optimization rule is given in equation (9.9) on page 134 together with explanations.

9.4.3 Experimental results

As already mentioned a GPC control without filtering of the measured values cannot be used properly; hence, a controller structure with integrated filter characteristics is selected. The prediction horizons are $N_p = 3$ and $N_u = 2$; the degrees of the system polynomials are $na = 1$, $nb = 1$ and $nt = 1$. With this simple model (figure 9.6) it is not possible to consider disturbances and hence, this is not implemented either. The following system and polynomial matrices result:

$$\mathbf{G}(z^{-1}) = \begin{bmatrix} \frac{0.1713\,z^{-1}}{1-0.9873\,z^{-1}} & -\frac{0.08566\,z^{-1}}{1-0.9873\,z^{-1}} & -\frac{0.08566\,z^{-1}}{1-0.9873\,z^{-1}} \\ 0 & \frac{0.1484\,z^{-1}}{1-0.9873\,z^{-1}} & -\frac{0.1484\,z^{-1}}{1-0.9873\,z^{-1}} \\ 0 & 0 & 0 \end{bmatrix}$$

$$\mathbf{T}(z^{-1}) = \begin{bmatrix} 1-0.95\,z^{-1} & 0 & 0 \\ 0 & 1-0.95\,z^{-1} & 0 \\ 0 & 0 & 1-0.95\,z^{-1} \end{bmatrix}$$

$$\mathbf{A}(z^{-1}) = \begin{bmatrix} 1-0.9873\,z^{-1} & 0 & 0 \\ 0 & 1-0.9873\,z^{-1} & 0 \\ 0 & 0 & 1 \end{bmatrix}$$

$$\mathbf{B}(z^{-1}) = \begin{bmatrix} 0.1713\,z^{-1} & -0.08566\,z^{-1} & -0.08566\,z^{-1} \\ 0 & 0.1484\,z^{-1} & -0.1484\,z^{-1} \\ 0 & 0 & 0 \end{bmatrix}$$

In order to compare the influence of the weighting of the actuating variables, two different matrices were selected as λ, one for a light weighting of the change of the values of the actuating variables with $\lambda = 0.001$ and one for a significantly higher weighting of the change with $\lambda = 0.1$.

Equation (9.9), derived from equation (8.14), shows that for the weighting not the values of the actuating variables themselves are considered, but the *change of the values of the actuating variables.* A change in one of the three inputs a, b or c means that a half bridge has to be switched. Hence, with the weighting factor λ the number of inverter switching actions is indeed evaluated (see chapter 9.3).

For reasons explained in chapter 9.2, a DMPC controller is better realized in stationary coordinates than in a rotating reference frame. Hence, results shown in $\alpha\beta$ coordinates are more meaningful. Figure 9.7 shows the behavior of a simple MIMO-DMPC current controller in stationary state. Matching the explanations in chapter 9.2, no contouring error occurs.

Comparing figure 9.7(a) with figure 9.7(b) shows the influence of the weighting factor λ on the behavior of the closed control loop. A higher value for λ leads to a higher weighting of the change of the actuating variables; hence, the controller now accepts a higher control error in order to avoid frequent switch-

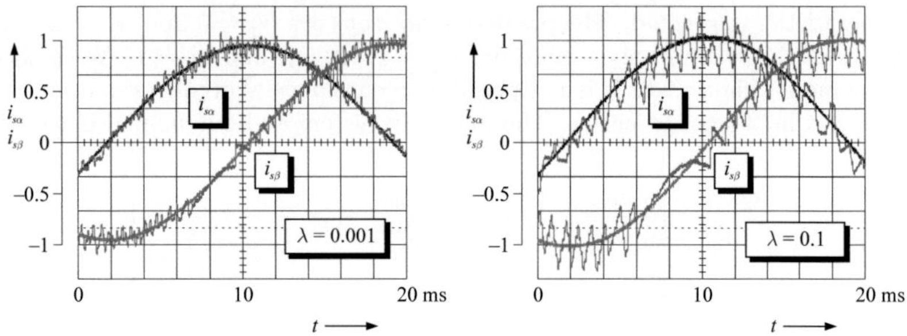

(a) $\lambda = 0.001$ (b) $\lambda = 0.1$

Figure 9.7: Current control with DMPC controller, stationary operation

ing of the semiconductors. It seems that similar results can be obtained when using a simple hysteresis controller—however, this is not true because there are some serious differences:

- When using a hysteresis controller, a switching event only takes place when a constant error limit has been exceeded. However, DMPC selects an optimal future switching sequence in advance according to the selected optimization criteria. This means that dependent on the current system state, switching takes place at different error values.
- A hysteresis controller does not really evaluate the switching effort. Thus a hysteresis-based two-level controller can e. g. not decide which one of the two zero voltage space vectors, being able to be switched by the inverter, is favorable for less switching actions.
- When using a hysteresis controller, it is not possible to exclude certain undesired switching states of the inverter already in the beginning of the optimization (see chapter 9.2.1).

Furthermore, it is to be noted that the weighting factor does influence the number of switching events and the control error, but not the dynamics of the entire system. Figure 9.8 showing the step response of the controlled system confirms this. In order to evaluate the mutual influence of $i_{s\alpha}$ and $i_{s\beta}$ besides the system dynamics, only the reference value for $i_{s\alpha}$ was step changed. As it can be seen easily, the rise time in figure 9.8(a) with $\lambda = 0.001$ as well as with $\lambda = 0.1$ in figure 9.8(b) is exactly the same and corresponds to the physically possible minimum value. Only the number of switching events is much higher when less importance is given to the switching effort (figure 9.8(a)). A mutual influence of both stator current components cannot be detected since the amplitude of the current harmonics is significantly higher than the influence of the cross coupling. Hence, it makes no sense to add a consideration of the cross coupling between $i_{s\alpha}$ and $i_{s\beta}$ to the simple machine model according to figure 9.6 because it would only result in a higher controller complexity without an improvement of the control quality.

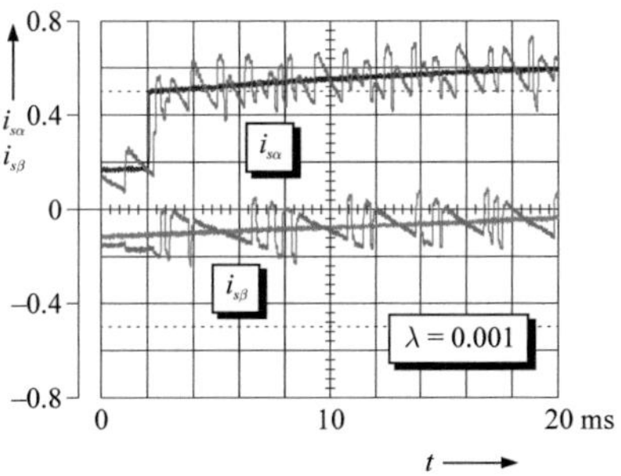

(a) $\lambda = 0.001$

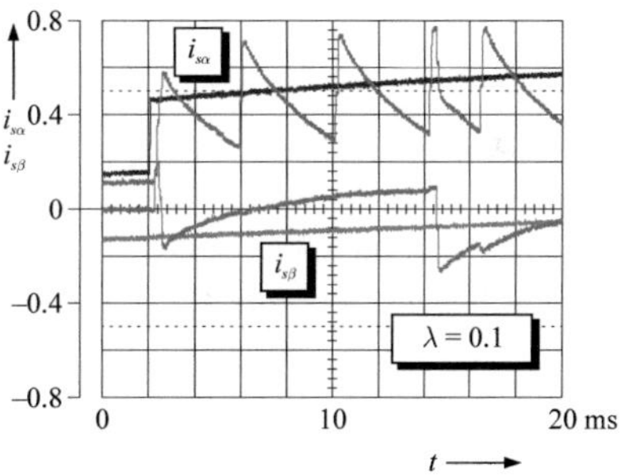

(b) $\lambda = 0.1$

Figure 9.8: Current control with DMPC controller, dynamic operation

9.4.4 Computation times

For investigating the feasibility of DMPC for drive systems, the computation times for different controller configurations and solution methods were determined. The corresponding results can be seen in table 9.2.

As it can be seen easily, considering methods with full enumeration control horizons N_u of more than two prediction steps leads to unacceptable computation times. Thereby, not just the higher number of possible switching sequences becomes noticeable; enlarging the horizons enlarges the different matrices further, which leads to an increase of the computational effort necessary for the calculation of the cost function J. Since, considering complete enumeration, the cost function J has to be evaluated for *every possible sequence of switching states*, both increases of the computational effort multiply with each other.

Method	N_u	Events	Max. computation time
Complete Enumeration	2	64	35 µs
Complete Enumeration	3	512	> 500 µs
Branch and Bound	2	$\leqq 64$	27 µs
Branch and Bound	3	$\leqq 512$	186 µs

Table 9.2: Computation times for a DMPC controller

Thus, a DMPC current control with a control horizon of $N_u = 3$ is only feasible with sampling cycles above 650 µs, corresponding to a maximum sampling frequency of 1.5 kHz. Control horizons with $N_u > 3$ cannot be realized due to the low sampling frequency needed.

The application of the Branch and Bound method for current control of an induction machine shows that, with smaller control horizons, the reduction of the computational effort becomes, as expected, only slightly noticeable. Taking the same basic conditions as for exhaustive search, for $N_u = 2$ a maximum computation time of 27 µs results. For a control horizon of $N_u = 3$ the computation lasts between 4.6 µs and 186 µs. The big difference between the maximum and minimum computation time is typical for Branch and Bound methods because, dependent on the current system state and the reference values, the optimum switching sequence is found in a different amount of time. Nevertheless, the experiments show that even with Branch and Bound, a DMPC control with control horizons higher than $N_u = 2$ is not suitable for drive control.

The comparison between complete enumeration and a search strategy based on Branch and Bound shows that by using an intelligent search algorithm, a reduction of the computation time is possible. This acceleration, however, is absolutely not sufficient for the needs in drive systems. However, from mathematics, other very elegant solution algorithms, which can save much computation time, are known especially for boolean optimization problems. As an example, one can mention the method by Laserre for quadratic boolean optimization tasks [78]. Solution strategies for linear boolean programs are discussed by Terno [115, chapter 5.10]. As the mathematical research in this area is not completed, providing an entire list of boolean optimization algorithms here does not make sense; readers interested in mathematics can look for papers with the AMS classification *90C09*[1] in adequate data bases. Indeed, it is questionable if the methods presented there will lead to success, i. e. to a

[1] The number 90C09 is the *American Mathematical Society* code number for "Boolean Programming".

sufficiently fast solution of the optimization task. Even the application of very specialized search strategies won't reduce the computation time necessary for the online-solution of the optimization problem to a value which is needed for a reasonable use of the method in drive technology. Hence, direct model-based predictive control with an *implicit solution* of the optimization problem is *not feasible* in drive technology.

9.5 Explicit solution

Contrary to most researchers who deal with model-based predictive control and who solve the emerging quadratic or linear optimization problem implicitly, Bemporad and Morari have chosen another way: They try to solve the optimization problem in dependence of the state vector $\mathbf{x}$ of the system explicitly, i. e. quasi offline [11, 12]. In contrast to the methods described so far, which are based on GPC and which make use of a transfer function based system model, the classical MPC approach used by Bemporad and Morari uses a state space model of the system to be controlled (compare equations (5.4) and (5.5) on page 32):

$$\mathbf{x}(t+1) = \mathbf{A}\mathbf{x}(t) + \mathbf{B}\mathbf{u}(t) \tag{9.10}$$

$$\mathbf{y}(t) = \mathbf{C}\mathbf{x}(t) + \mathbf{D}\mathbf{u}(t) \tag{9.11}$$

under consideration of the constraints

$$\mathbf{x} \in \boldsymbol{X} \subset \mathrm{IR}^n, \qquad \mathbf{y} \in \boldsymbol{Y} \subset \mathrm{IR}^p, \qquad \mathbf{u} \in \boldsymbol{U} \subset \mathrm{IR}^m$$

Furthermore, a cost function is defined as follows:

$$\begin{aligned} J(N_1, N_2) = {} & \mathbf{x}(t+N_2)^T \, \mathbf{P} \, \mathbf{x}(t+N_2) \\ & + \sum_{j=N_1}^{N_2} \Big(\mathbf{x}(t+j-1)^T \, \mathbf{Q} \, \mathbf{x}(t+j-1) \\ & \qquad + \mathbf{u}(t+j-1)^T \, \mathbf{R} \, \mathbf{u}(t+j-1) \Big) \end{aligned} \tag{9.12}$$

In this case we have a quadratic cost function with finite time horizon, $\mathbf{Q}$ weights the system states and $\mathbf{R}$ the actuating variables; $\mathbf{P}$ weights the final value of the state vector. Linear cost functions with a finite or infinite cost horizon are also possible. Furthermore, the cost function (9.12) has been simplified compared to (6.6) and (8.14) by assuming the reference value $\mathbf{w}(t+j)$

being zero for all $j \geqq 0$. As already known, $N_1 = 1$ and $N_2 = N_p$ are chosen for reasons of simplicity. Now, this cost function has to be minimized under the following constraints:

$$
\begin{array}{ll}
\mathbf{x}(t+j) \in \boldsymbol{X} & j = 1 \dots N_p \\
\mathbf{u}(t+j) \in \boldsymbol{U} & j = 0 \dots N_p \\
\mathbf{x}(t+j+1) = \mathbf{A}\mathbf{x}(t+j) + \mathbf{B}\mathbf{u}(t+j) & k \geqq 0 \\
\mathbf{u}(t+j) = \mathbf{K}\mathbf{x}(t+j) & N_1 \leqq k \leqq N_2 \\
\mathbf{Q} = \mathbf{Q}^T \succeq 0 & (\text{“}\succeq\text{” means positive semidefinite}) \\
\mathbf{R} = \mathbf{R}^T \succ 0 & (\text{“}\succ\text{” means positive definite}) \\
\mathbf{P} \succeq 0 &
\end{array}
$$

By replacing

$$\mathbf{x}(t+j) = \mathbf{A}^j \mathbf{x}(t) + \sum_{k=0}^{j-1} \mathbf{A}^k \mathbf{B}\mathbf{u}(t+j-1-k)$$

the optimization problem can be obtained, i. e. to minimize the cost function (9.12), in the form:

$$J'\big(\mathbf{x}(t)\big) = \frac{1}{2}\mathbf{x}(t)^T \, \mathbf{Y} \, \mathbf{x}(t) + \min_{\mathbf{U}} \left\{ \frac{1}{2}\mathbf{U}^T \mathbf{H} \mathbf{U} + \mathbf{x}(t)^T \mathbf{F} \mathbf{U} \right\} \tag{9.13}$$

with the constraints

$$\mathbf{G}\mathbf{U} \leqq \mathbf{W} + \mathbf{E}\mathbf{x}(t)$$

with

$$
\begin{array}{ll}
\mathbf{U} = \left[\mathbf{u}(t)^T, \dots, \mathbf{u}(t+N_u-1)^T\right]^T \in \mathbb{R}^s & s = m \cdot N_u \\
\mathbf{H} = \mathbf{H}^T \succ 0 &
\end{array}
$$

The Matrices **H**, **F**, **Y**, **G**, **W** and **E** can be calculated from **Q**, **R** and from the original optimization rules, based on (9.12). Since only the vector of the actuating variables **U** is required as a result of the optimization, the term with **Y** in (9.13) can be neglected.

The optimization problem in (9.13) is a so-called *Quadratic Program (QP)*. Since J' is dependent on the current system state $\mathbf{x}(t)$, the QP has to be solved again for every sampling cycle. Although time efficient solution algorithms for

QP problems are known today, the online calculation of $\mathbf{u}(t)$ for fast applications in control engineering is hardly feasible, especially for the fast processes in drive technology. Hence, the application of the "classical" MPC method will be limited to processes with low sampling rate and/or low complexity.

In [11], Bemporad, Morari et al. proposed a new method for linear time-invariant (LTI) systems which transfer the solution of the optimization problem *offline*; in [12] this strategy is expanded to piecewise affine (PWA) and hybrid systems. The basic idea is to treat the state vector $\mathbf{x}(t)$ as a parameter vector. Thereby, the optimization problem (9.13) becomes a *multi-parametric Quadratic Program*, short *mp-QP*. By solving this mp-QP problem, one obtains a solution function $\mathbf{U}_{opt} = f\big(\mathbf{x}(t)\big)$; hence, the MPC control law is explicitly available. In the following, it is described how this solution function can be obtained and which characteristics it has.

First, an auxiliary variable is defined:

$$\mathbf{z} = \mathbf{U} + \mathbf{H}^{-1}\mathbf{F}^T\mathbf{x}(t) \qquad \mathbf{z} \in \mathrm{I\!R}^s \tag{9.14}$$

By quadratic complement, (9.13) is transformed to

$$J'_z\big(\mathbf{x}(t)\big) = \min_{\mathbf{z}} \left\{ \frac{1}{2}\mathbf{z}^T\mathbf{H}\mathbf{z} \right\} \tag{9.15}$$

under the constraint

$$\mathbf{G}\mathbf{z} \leqq \mathbf{W} + \mathbf{S}\mathbf{x}(t)$$

with

$$\mathbf{S} = \mathbf{E} + \mathbf{G}\mathbf{H}^{-1}\mathbf{F}^T$$

$$J'_z\big(\mathbf{x}(t)\big) = J'\big(\mathbf{x}(t)\big) - \frac{1}{2}\mathbf{x}^T\left(\mathbf{Y} - \mathbf{F}\mathbf{H}^{-1}\mathbf{F}^T\right)\mathbf{x}(t)$$

Thus, the state vector $\mathbf{x}(t)$, considered as a parameter, only appears in the constraints.

In order to solve the mp-QP problem, an initial vector $\mathbf{x}_0$ within the possible state space, for which the upper optimization problem can be solved, has to be found now. How such a start point can be found is described e. g. by Bemporad et al. [13]. With the help of $\mathbf{x}_0$, obtained as described above, an optimum solution $\mathbf{z}_0$ of (9.15) can be determined. Since the Matrix $\mathbf{H}$ is positive definite, the solution is unique and a set of active constraints $\tilde{\mathbf{G}}\mathbf{z}_0 = \tilde{\mathbf{W}} + \tilde{\mathbf{S}}\mathbf{x}_0$ is obtained. These constraints form a region $\mathcal{P}_0$. The optimum value of $\mathbf{z}$ and the corresponding vector of the lagrange multiplier λ are now uniquely defined affine functions of $\mathbf{x}$ over $\mathcal{P}_0$; this can be proved with the help of the Karush-Kuhn-Tucker conditions. In the same way, it can be proved that $\mathcal{P}_0$ forms a polyhedron or a polytope in state space.

If the first region $\mathcal{P}_0$ is defined, the remaining state space has to be analyzed and further regions have to be created. A possible approach is also described by Bemporad et al. in [13]. Hence, as the issues proven for $\mathcal{P}_0$ are also valid for all regions, these are also polytopes, each with an affine law for determining an optimum $\mathbf{z}$. As the relationship between $\mathbf{z}$ and $\mathbf{U}$ is also affine (see equation (9.14)), a piecewise affine description $\mathbf{U}_{opt} = f(\mathbf{x}(t))$ exists now.

Indeed, for users less interested in mathematics, it is not necessary to deal in detail with the algorithms and mathematical proofs mentioned and also partly explained above. At the Swiss Federal Institute of Technology (ETH) Zurich at Prof. Morari's department, a toolbox for MATLAB was developed with which calculation of explicit solutions for different LTI and PWA systems with different cost functions can easily be done [73]. Figure 9.9 shows the results of a calculation of the explicit solution for current control of an induction machine. Here, a 1-norm cost function with a prediction horizon of two steps was selected. For an easier rating of the obtained piecewise affine solution function, the solution was determined for the origin as fixed reference value. As it can easily be seen, the three phases a, b, and c are always switched in such a way that the current space vector is moved to the origin. In the origin itself the vector $[0, 0, 0]$ is selected, which leads to the actual value of the current space vector remaining on this point.

The trajectories of a closed-loop control using a controller with a PWA control law are shown in figure 9.10. As an example, a closer look is taken at two trajectories in the following.

Considering trajectory A at the beginning, at point A_1, as it can be obtained from a comparison with figure 9.9, the state $[1, 0, 0]$ is switched, i. e. $c = 1$, $b = 0$ and $a = 0$. This voltage space vector moves the system state in the fastest way towards the origin although the origin cannot be reached directly. At point A_2 the state vector $\mathbf{x}$ reaches the value $i_{s\beta} = 0$; hence, according to the predetermined control law, a switching operation takes place in phase b (see figure 9.9(b)). Now, the switching state $[1, 1, 0]$ is applied to the machine which moves the state vector directly into the origin along the axis $i_{s\beta} = 0$. Then, the half bridges b and c are again switched back to the "lower" semiconductors, leading to a zero vector applied to the machine. Hence, the system state will remain in the origin.

If trajectory B is considered, first of all, in point B_1, the state $[1, 0, 0]$ is applied to the machine, too. Five sampling cycles later, the system state reaches

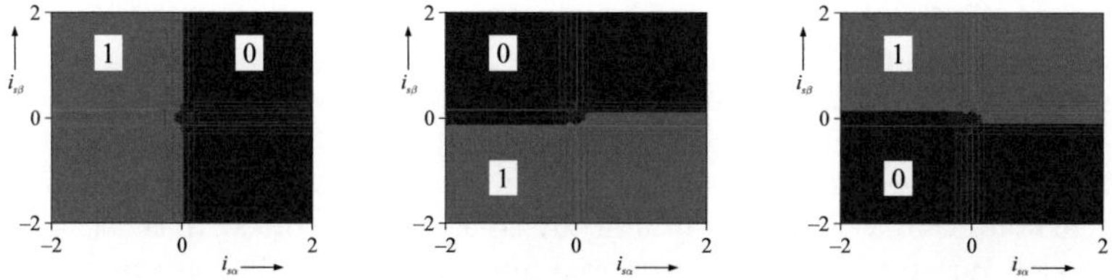

(a) Phase a (b) Phase b (c) Phase c

Figure 9.9: Explicit solution for three half bridges

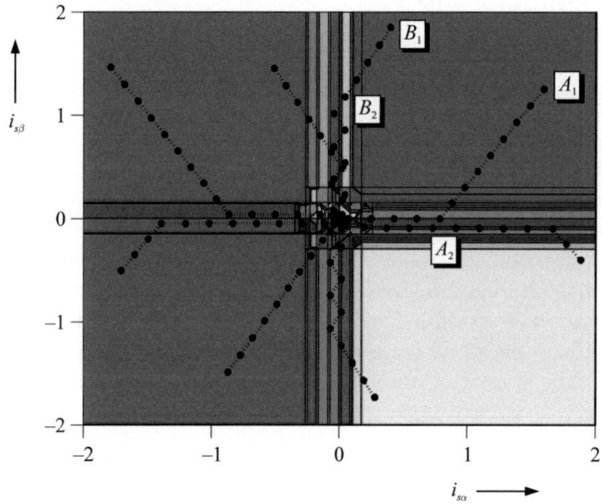

Figure 9.10: Trajectories of the explicit solution

the value $i_{s\alpha} = 0$. Since there is no voltage vector being switchable by the inverter that can move $\mathbf{x}$ directly into the origin along the axis $i_{s\alpha} = 0$, the system state is oscillating between the sectors for $a = 1$ and $a = 0$ (figure 9.9(a)). According to that, the states $[1, 0, 1]$ and $[1, 0, 0]$ are alternately applied to the machine. Thus, the machine state is forced to slide along the boundary between the sectors with the control law $\mathbf{u} = [1, 0, 1]$ and $\mathbf{u} = [1, 0, 0]$. Here the DMPC method gains the character of sliding mode control [35], because the system state is moved along synthetically created trajectories. In this case, the controlled variable $\mathbf{x}$ will also be moved into the origin as fast as possible.

The determined explicit solution consists of 217 polytopes (regions), each with a piecewise affine control law being valid. At first view, this seems to be remarkable since a two-level inverter has only eight different switching states and hence for the next step only eight different control laws can exist. However, it has to be noted that the solution vector $\mathbf{U}$ is determined over the whole prediction horizon when the explicit solution is calculated. Hence, in the case treated here already $8^2 = 64$ different control laws are possible. Furthermore, the complete number of determined regions can be significantly higher than the maximum number of control laws being theoretically possible because of the constraint that all regions have to be convex polytopes.

Tracking

For the practical application of a control system, the optimization to a fixed reference value described so far can rarely be used because in practical operation, the reference values will change dynamically. Correspondingly, the plant model has to be changed so that not the system state $\mathbf{x}$ itself, but the control error $\mathbf{x} - \mathbf{w}$ has to be controlled to become zero. This procedure in which the actual value has to follow a "free" reference value is called *tracking*. For that, the state vector $\mathbf{x}$ is extended by the actuating variable $\mathbf{u}$ and by the reference value $\mathbf{w}$, and as input variable the change of the values of the actuating variable, i. e. $\Delta\mathbf{u}$, is used. Consequently, the model equation (9.10) becomes

$$\begin{bmatrix} \mathbf{x}(t+1) \\ \mathbf{u}(t) \\ \mathbf{w}(t+1) \end{bmatrix} = \begin{bmatrix} \mathbf{A} & \mathbf{B} & \mathbf{0} \\ \mathbf{0} & \mathbf{I} & \mathbf{0} \\ \mathbf{0} & \mathbf{0} & \mathbf{I} \end{bmatrix} \cdot \begin{bmatrix} \mathbf{x}(t) \\ \mathbf{u}(t-1) \\ \mathbf{w}(t) \end{bmatrix} + \begin{bmatrix} \mathbf{B} \\ \mathbf{I} \\ \mathbf{0} \end{bmatrix} \cdot \Delta\mathbf{u}(t)$$

Thus, the new model equations can be obtained:

$$\mathbf{x}_t(t+1) = \mathbf{A}_t\mathbf{x}_t(t) + \mathbf{B}_t\mathbf{u}_t(t) \tag{9.16}$$

$$\mathbf{y}_t(t) = \mathbf{C}_t\mathbf{x}_t(t) + \mathbf{D}_t\mathbf{u}_t(t) \tag{9.17}$$

with

$$\mathbf{x}_t(t) = \begin{bmatrix} \mathbf{x}(t+1) \\ \mathbf{u}(t) \\ \mathbf{w}(t+1) \end{bmatrix}, \qquad \mathbf{u}_t(t) = \Delta\mathbf{u}(t) ,$$

$$\mathbf{A}_t = \begin{bmatrix} \mathbf{A} & \mathbf{B} & \mathbf{0} \\ \mathbf{0} & \mathbf{I} & \mathbf{0} \\ \mathbf{0} & \mathbf{0} & \mathbf{I} \end{bmatrix}, \qquad \mathbf{B}_t = \begin{bmatrix} \mathbf{B} \\ \mathbf{I} \\ \mathbf{0} \end{bmatrix},$$

$$\mathbf{C}_t = \mathbf{I}, \qquad = \mathbf{D}_t = \mathbf{0}$$

The cost function known from equation (9.12)

$$J = \sum_{j=N_1}^{N_2} \Big(\mathbf{x}(t+j-1)^T \, \mathbf{Q} \, \mathbf{x}(t+j-1) + \mathbf{u}(t+j-1)^T \, \mathbf{R} \, \mathbf{u}(t+j-1) \Big)$$

is also modified so that it contains the evaluation of the control error $\mathbf{x} - \mathbf{w}$:

$$J = \sum_{j=N_1}^{N_2} \left(\begin{bmatrix} \mathbf{x}(t+j) \\ \mathbf{u}(t+j-1) \\ \mathbf{w}(t+j) \end{bmatrix}^T \cdot \begin{bmatrix} \mathbf{Q} & \mathbf{0} & -\mathbf{Q} \\ \mathbf{0} & \mathrm{R} & \mathbf{0} \\ -\mathbf{Q} & \mathbf{0} & \mathbf{Q} \end{bmatrix} \cdot \begin{bmatrix} \mathbf{x}(t+j) \\ \mathbf{u}(t+j-1) \\ \mathbf{w}(t+j) \end{bmatrix} \right.$$
$$\left. + \, \Delta\mathbf{u}(t+j-1)^T \cdot \mathbf{R}_\Delta \cdot \Delta\mathbf{u}(t+j-1) \right)$$

The new weighting factor $\mathbf{R}_\Delta$ thereby weights the change of the values of the actuating variables $\Delta\mathbf{u}$. Now, adequate to the new plant model (9.16)-(9.17), a cost function

$$J = \sum_{j=N_1}^{N_2} \Big(\mathbf{x}_t(t+j-1)^T \, \mathbf{Q}_t \, \mathbf{x}_t(t+j-1) + \mathbf{u}_t(t+j-1)^T \, \mathbf{R}_t \, \mathbf{u}_t(t+j-1) \Big) \tag{9.18}$$

with

$$\mathbf{Q}_t = \begin{bmatrix} \mathbf{Q} & \mathbf{0} & -\mathbf{Q} \\ \mathbf{0} & \mathrm{R} & \mathbf{0} \\ -\mathbf{Q} & \mathbf{0} & \mathbf{Q} \end{bmatrix}, \qquad \mathbf{R}_t = \mathbf{R}_\Delta$$

exists.

If, instead of the model given in the equations (9.10) and (9.11) and the cost function (9.12), the model in (9.16) and (9.17) and the cost function (9.18) is used and one proceeds as described so far in order to determine optimal solutions, an explicit solution of the MPC problem under consideration of reference values is possible. Due to the fact that the dimension of the state vector enlarges when reference values and former values of the actuating variables are considered, the dimension of the solution space will be enlarged accordingly. Hence, the solution of a tracking problem is always more complex than a solution which takes a zero vector as constant reference value.

Delays

Normally, digital controls cause a delay of one sampling cycle between the point of time at which the controller has determined a reference value and the point of time at which this reference value is actually applied to the plant. MPC strategies based on a transfer function-based model of the plant can easily be adopted in this case because the delay can be implemented easily by multiplying the polynomial $B(z^{-1})$ or the polynomial matrix $\mathbf{B}(z^{-1})$ of the CARIMA model with z^{-1}, as described in chapter 6.1.1 on page 39. The use of a state space-based predictive controller unfortunately excludes this simple method; however, it is possible to add an appropriate delay to the plant model by extending the state vector. Thus, the state equation of the system is modified as follows:

$$\begin{bmatrix} \mathbf{x}(t+1) \\ \mathbf{u}(t) \end{bmatrix} = \begin{bmatrix} \mathbf{A} & \mathbf{B} \\ \mathbf{0} & \mathbf{0} \end{bmatrix} \cdot \begin{bmatrix} \mathbf{x}(t) \\ \mathbf{u}(t-1) \end{bmatrix} + \begin{bmatrix} \mathbf{0} \\ \mathbf{I} \end{bmatrix} \cdot \mathbf{u}(t)$$

Consequently, a new representation of the model can be obtained:

$$\mathbf{x}_d(t+1) = \mathbf{A}_d \mathbf{x}_d(t) + \mathbf{B}_d \mathbf{u}_d(t) \tag{9.19}$$

$$\mathbf{y}_d(t) = \mathbf{C}_d \mathbf{x}_d(t) + \mathbf{D}_d \mathbf{u}_d(t) \tag{9.20}$$

with

$$\mathbf{x}_d(t) = \begin{bmatrix} \mathbf{x}(t+1) \\ \mathbf{u}(t) \end{bmatrix}, \qquad \mathbf{u}_d(t) = \mathbf{u}(t)\,,$$

$$\mathbf{A}_d = \begin{bmatrix} \mathbf{A} & \mathbf{B} \\ \mathbf{0} & \mathbf{0} \end{bmatrix}, \qquad \mathbf{B}_d = \begin{bmatrix} \mathbf{0} \\ \mathbf{I} \end{bmatrix},$$

$$\mathbf{C}_d = \mathbf{I}\,, \qquad \mathbf{D}_d = \mathbf{0}$$

The vector of state variables $\mathbf{x}_d$ does now also contain the actuating variable $\mathbf{u}$. In the cost function, however, only the state variables themselves should be evaluated. Thus equation (9.12) has to be adjusted again. In this case, it can be applied:

$$J = \sum_{j=N_1}^{N_2} \left(\begin{bmatrix} \mathbf{x}(t+j) \\ \mathbf{u}(t+j-1) \end{bmatrix}^T \cdot \begin{bmatrix} \mathbf{Q} & \mathbf{0} \\ \mathbf{0} & \mathbf{0} \end{bmatrix} \cdot \begin{bmatrix} \mathbf{x}(t+j) \\ \mathbf{u}(t+j-1) \end{bmatrix} + \mathbf{u}(t+j-1)^T \cdot \mathbf{R} \cdot \mathbf{u}(t+j-1) \right)$$

Here not the *change* of the values of the actuating variable, but again the actuating variable itself is evaluated. Thus, the new cost function can be written as:

$$J = \sum_{j=N_1}^{N_2} \left(\mathbf{x}_d(t+j-1)^T \, \mathbf{Q}_d \, \mathbf{x}_d(t+j-1) + \mathbf{u}_d(t+j-1)^T \, \mathbf{R}_d \, \mathbf{u}_d(t+j-1) \right) \tag{9.21}$$

with

$$\mathbf{Q}_d = \begin{bmatrix} \mathbf{Q} & \mathbf{0} \\ \mathbf{0} & \mathbf{0} \end{bmatrix}, \qquad \mathbf{R}_d = \mathbf{R}$$

If the state equations (9.19) and (9.20) and the corresponding cost function (9.21) are used and the problem is solved via mp-QP, an explicit solution of the optimization problem is obtained for the case that the system to be controlled has a delay of one sampling cycle.

Delay with tracking

It can be assumed that drive controls commonly have a delay and that the system state should not be driven into the origin, but it should follow a reference value. Thus, for an applicable control for electrical drives both methods mentioned above have to be combined. This can be achieved by a further slight modification of (9.16),

$$\begin{bmatrix} \mathbf{x}(t+1) \\ \mathbf{u}(t) \\ \mathbf{w}(t+1) \end{bmatrix} = \begin{bmatrix} \mathbf{A} & \mathbf{B} & \mathbf{0} \\ \mathbf{0} & \mathbf{I} & \mathbf{0} \\ \mathbf{0} & \mathbf{0} & \mathbf{I} \end{bmatrix} \cdot \begin{bmatrix} \mathbf{x}(t) \\ \mathbf{u}(t-1) \\ \mathbf{w}(t) \end{bmatrix} + \begin{bmatrix} \mathbf{0} \\ \mathbf{I} \\ \mathbf{0} \end{bmatrix} \cdot \Delta\mathbf{u}(t)$$

which leads to the model equations

$$\mathbf{x}_{t,d}(t+1) = \mathbf{A}_{t,d}\mathbf{x}_{t,d}(t) + \mathbf{B}_{t,d}\mathbf{u}_{t,d}(t) \tag{9.22}$$

$$\mathbf{y}_{t,d}(t) = \mathbf{C}_{t,d}\mathbf{x}_{t,d}(t) + \mathbf{D}_{t,d}\mathbf{u}_{t,d}(t) \tag{9.23}$$

with

$$\mathbf{x}_{t,d}(t) = \begin{bmatrix} \mathbf{x}(t+1) \\ \mathbf{u}(t) \\ \mathbf{w}(t+1) \end{bmatrix} , \qquad \mathbf{u}_{t,d}(t) = \Delta\mathbf{u}(t) ,$$

$$\mathbf{A}_{t,d} = \begin{bmatrix} \mathbf{A} & \mathbf{B} & \mathbf{0} \\ \mathbf{0} & \mathbf{I} & \mathbf{0} \\ \mathbf{0} & \mathbf{0} & \mathbf{I} \end{bmatrix} , \qquad \mathbf{B}_{t,d} = \begin{bmatrix} \mathbf{0} \\ \mathbf{I} \\ \mathbf{0} \end{bmatrix} ,$$

$$\mathbf{C}_{t,d} = \mathbf{I} , \qquad \mathbf{D}_{t,d} = \mathbf{0}$$

The evaluation function (9.18) does not have to be modified, i. e. $\mathbf{Q}_{t,d} = \mathbf{Q}_t$ and $\mathbf{R}_{t,d} = \mathbf{R}_t$, in which again the weighting factor $\mathbf{R}_\Delta$ weights the change of the value of the actuating variables.

As the above equations show, for the integration of a delay of one sampling cycle into a control with tracking no extension of the state vector compared to the mere tracking version $\mathbf{x}_t$ is necessary. Hence, the consideration of the delay can, in this case, be integrated into the plant model without any further effort.

9.5.1 Standard algorithm

If the explicit solution of MPC is calculated with the method described above, for a practical realization of the control one first has to find out in which one of the calculated polytopes the current system state $\mathbf{x}$ lies in order to apply the corresponding affine control law afterwards. As the polytopes are usually not sorted there is no other way than an exhaustive search over all existing polytopes. For every polytope $\mathcal{P}_i$ and for all system states $\mathbf{x}$ being within this polytope the equation

$$\mathbf{H}_i\mathbf{x} \leqq \mathbf{K}_i \tag{9.24}$$

is valid. Then all determined polytopes have to be checked one after another if equation (9.24) is valid. If this is the case, the polytope $\mathcal{P}_j$ that contains $\mathbf{x}$ is found. Then the corresponding control law has to be applied.

Due to the complete enumeration of all existent polytopes this method is, of course, not very efficient. Thus, more intelligent search strategies are recommended for drive control.

9.5.2 Minimum-Time Controller

The so-called *Minimum-Time Controller* proposed by Grieder/Morari is considerably faster than the standard method. It can be used for linear time-invariant

(LTI) [45] as well as for piecewise affine (PWA) or hybrid systems [46]. The intention of this method is to determine an indeed sub-optimal, but in the execution less complex solution of the optimization task. A lower complexity of the explicit controller structure can in the easiest way be obtained with short prediction horizons and with a lesser number of input variables. Unfortunately, these variables are normally dependent of the system to be controlled and thus cannot be chosen freely. The alternative way proposed here consists of the iterative solution of single-step optimization problems with a varying target region $\mathcal{T}_{set}$. For that in each case the following optimization task

$$J_1\big(\mathbf{x}(t)\big) = \min_{\mathbf{u}(t)} \Big\{ \mathbf{x}(t+1)^T \, \mathbf{Q} \, \mathbf{x}(t+1) + \mathbf{u}(t)^T \, \mathbf{R} \, \mathbf{u}(t) \Big\} \tag{9.25}$$

is solved under consideration of

$$\begin{aligned} \mathbf{x}(t+1) &\in \mathcal{T}_{set} \subseteq \boldsymbol{X} \\ \mathbf{u}(t) &\in \boldsymbol{U} \\ \mathbf{x}(t+1) &= \mathbf{A}\mathbf{x}(t) + \mathbf{B}\mathbf{u}(t) \\ \mathbf{Q} &= \mathbf{Q}^T \succ 0 \\ \mathbf{R} &= \mathbf{R}^T \succ 0 \end{aligned}$$

The intention is to drive the system to be controlled as fast as possible into a region $\mathcal{X}_{LQR}$ in which an optimal unconstrained LQR control is feasible without exceeding the constraints. This set shall be the first target region:

$$\mathcal{T}_{set} = \left(\mathcal{X}_f^1\right)^0 = \mathcal{X}_{LQR}$$

Within this region the LQR control law $\mathbf{F}_{LQR}$ is valid so that $\mathbf{F}_0^0 = \mathbf{F}_{LQR}$ and $\mathbf{G}_0^0 = 0$.

In order to obtain a PWA control law for the whole state space, the mp-QP based on (9.25) is solved. This results in the following polytope structure:

$$\mathcal{P}_k^{iter} = \Big\{ \mathbf{x} \in \mathbb{R}^n | \mathbf{H}_k^{iter} \mathbf{x} \leqq \mathbf{K}_k^{iter} \Big\} \qquad k = 1 \dots R$$

In this case, *iter* is the iteration step, beginning from 0 for $\mathcal{X}_{LQR}$; R is the number of different polytopes in the current iteration step. The control laws belonging to the individual polytopes can be expressed by the equation

$$\mathbf{u}(t) = \mathbf{F}_k^{iter} \mathbf{x}(t) + \mathbf{G}_k^{iter}$$

Now, the explicit solution can be calculated iteratively for the whole region. For this purpose, the solution region of the last iteration step is selected as the new target region:

$$\mathcal{T}_{set} = \left(\mathcal{X}_f^1\right)^{iter} = \bigcup_{k=1}^{R} \mathcal{P}_k^{iter}$$

For this region, a single-step mp-QP problem is solved again whereby the new control step *iter + 1* results. This is repeated until the target region does not change anymore. After every iteration step, all regions and the corresponding control laws are saved.

Because of the iterative solution of the optimization task, the explicit solution gets a tree-like structure; this can be used for the setup of an efficient search strategy for the online evaluation of the PWA control law. Therefore, at first the region with the lowest iteration step containing the current system state $\mathbf{x}(t)$ is determined, i. e.

$$iter_{min} = \min_{iter} iter, \text{in which } \mathbf{x}(t) \in \left(\mathcal{X}_f^1\right)^{iter}$$

Within this region the active polytope $\mathcal{P}_j^{iter_{min}}$ is determined via exhaustive search. This can be done much faster than with direct exhaustive search because the number of polytopes per region is very small. If $iter$ and j are known, the value of the actuating variable to be applied to the plant can be determined with the equation

$$\mathbf{u}(t) = \mathbf{F}_j^{iter_{min}} \mathbf{x}(t) + \mathbf{G}_j^{iter_{min}} \tag{9.26}$$

The ulterior motive of the Minimum-Time Controller is, that, independent from the cost function, far away from the reference value, the maximum value of the actuating variable is in any case also the optimal one. This leads to a polytope structure simplified in two ways: On the one hand, the number of regions compared to the standard method is reduced; on the other hand, the individual polytopes and control laws are automatically sorted in a tree-like structure. The two-stage evaluation of the explicit control law significantly reduces the necessary online computational time compared to the standard method, however, for the cost of a non-optimum control concerning the cost function. Instead, the strategy is focused to move the system state $\mathbf{x}(t)$ as fast as possible into the origin or the region $\mathcal{X}_{LQR}$, respectively.

9.5.3 Binary search tree

Another method to reduce the online evaluation time is to transform the polytope structure obtained with the standard method into a binary search tree.

Tøndel, Johansen and Bemporad have proposed such a strategy in [117]. For this purpose, all hyperplanes, i. e. the boundaries of all polytopes, are considered in a first step. The whole number of hyperplanes shall be L and they are all represented by the equation $\mathbf{a}_j{}^T\mathbf{x} = \mathbf{b}_j$, $j = 1 \dots L$. With the help of this equation, a describing function $d_j(\mathbf{x})$ is defined as follows:

$$d_j(\mathbf{x}) = \mathbf{a}_j{}^T\mathbf{x} - \mathbf{b}_j \tag{9.27}$$

By evaluating the sign of $d_j(\mathbf{x})$ for every point in state space, it can be stated if the point is "above" or "below"[2] the corresponding hyperplane j. Consequently, any polytope $\mathcal{P}(\boldsymbol{J})$ in state space can be described by a so-called index set $\boldsymbol{J}$; an index set $\boldsymbol{J} = \{1^+, 2^-, 3^-\}$ would be equivalent to $d_1(\mathbf{x}) \geqq 0$, $d_2(\mathbf{x}) < 0$ and $d_3(\mathbf{x}) < 0$; it describes how the polytope $\mathcal{P}(\boldsymbol{J})$ behaves with each of the enumerated hyperplanes. Additionally, an index set $\boldsymbol{I}$ is defined that describes which polytopes $\mathcal{P}_i$ of the original polytope structure lie at least partly within the region described by $\mathcal{P}(\boldsymbol{J})$:

$$\boldsymbol{I}(\boldsymbol{J}) = \{i | \mathcal{P}_i \cap \mathcal{P}(\boldsymbol{J}) \text{ is full dimensional}\}$$

Furthermore, for every index set $\boldsymbol{I}$, a set $\boldsymbol{F}(\boldsymbol{I})$ exists which enumerates the control laws of the polytopes lying within $\mathcal{P}(\boldsymbol{J})$:

$$\boldsymbol{F}(\boldsymbol{I}) = \{k | F_k \text{ belonging to } \mathcal{P}_i, i \in \boldsymbol{I}\}$$

This last definition is important because different polytopes can have the same control law. This is particularly interesting considering Receding Horizon Control since, in this case, only the first value of the sequence of values of the actuating variables will be passed on to the system. Hence, for the construction of the search tree one only has to differentiate between the first elements of the control law which leads to a significantly simplified tree structure.

The concept of the method is to design a search tree in which for every node N_k for a given state vector $\mathbf{x}(t) \in \boldsymbol{X}$, the describing function $d_j(\mathbf{x})$ is called. Depending on the sign of $d_j(\mathbf{x})$, one will branch either in the following left or the right subtree. The indices of the nodes which do still have to be evaluated are stored in a list $\boldsymbol{K}$. For each of the nodes N_k, two index sets $\boldsymbol{J}_k$ exist, which contain the results of the describing functions $d_j(\mathbf{x})$ already applied and one set $\boldsymbol{I}_k$, which contains the indices of the polytopes contained in $\mathcal{P}(\boldsymbol{J}_k)$, i. e. $\boldsymbol{I}_k = \boldsymbol{I}(\boldsymbol{J}_k)$. Evaluated nodes additionally contain the information about

[2] The terms "above" and "below" are used in quotation marks because the definition of "above" and "below" in multidimensional spaces does not correspond to the natural feeling, but is only given by mathematics.

the affine control law F_k to be applied in this operating point if they are end nodes. Otherwise an array with the index of the hyperplane j_k whose describing function $d_{j_k}(\mathbf{x})$ is needed for the decision about the selection of the following part of the tree is added. The construction of this search tree is done in the following way:

At first, for all hyperplanes, the index sets $\boldsymbol{I}(j^+)$ and $\boldsymbol{I}(j^-)$ for $j = 1 \dots L$ are determined, i. e. the polytopes of the original description of the polytopes which lie at least partly "above" or "below" the concerning hyperplane have to be identified.

Afterwards, the root of the tree is initialized as node N_1. As the root encloses all available polytopes, it applies consequently that $\boldsymbol{J}_1$ is an empty set and $\boldsymbol{I}_1$ contains all polytopes. Since the root has not been evaluated yet, it is set as the first element in the set of the nodes which still have to be evaluated. Hence, $\boldsymbol{K} = \{N_1\}$. Then, the other nodes of the tree are determined iteratively. For this purpose, the following loop is run through until there are no more nodes to be evaluated, i. e. $\boldsymbol{K} = \emptyset$.

From the set $\boldsymbol{K}$, one node N_k is selected and $\boldsymbol{K}$ will become $\boldsymbol{K} \setminus N_k$. The polytopes $\mathcal{P}_i$ with $i \in \boldsymbol{I}_k$ being in the hyperplanes $\boldsymbol{J}_k$ which have already been evaluated are now separated again by all hyperplanes which have not been evaluated yet and the remaining regions are determined approximately with $\boldsymbol{I}(\boldsymbol{J}_k) \cap \boldsymbol{I}(j^\pm)$ for all $j = 1 \dots L$. The results of this separation are sorted by $\max(|\boldsymbol{F}(\boldsymbol{I}(\boldsymbol{J}_k) \cap \boldsymbol{I}(j^+))|, |\boldsymbol{F}(\boldsymbol{I}(\boldsymbol{J}_k) \cap \boldsymbol{I}(j^-))|)$. Then, for the first n_j elements of this sorted list, the exact index sets $\boldsymbol{I}_k^\pm = \boldsymbol{I}(\boldsymbol{J}_k \cup j^\pm)$ are calculated. The value of n_j is selected in such a way that all hyperplanes of the sorted list which minimize the number of the maximum remaining control laws, i. e. $\max(|\boldsymbol{F}(\boldsymbol{I}(\boldsymbol{J}_k) \cap \boldsymbol{I}(j^+))|, |\boldsymbol{F}(\boldsymbol{I}(\boldsymbol{J}_k) \cap \boldsymbol{I}(j^-))|)$, are used for the calculation again. This repeated calculation is necessary since the estimation of the remaining regions with $\boldsymbol{I}(\boldsymbol{J}_k) \cap \boldsymbol{I}(j^\pm)$ can result in a larger polytope than the mathematically exact, but also more complex method with $\boldsymbol{I}_k^\pm = \boldsymbol{I}(\boldsymbol{J}_k \cup j^\pm)$; the following example shall serve as an illustration of this fact.

When a polytope structure according to figure 9.11 is given and only the polytope $\mathcal{P}_1$ and both hyperplanes j_1 and j_2 are considered, apparently the following statements are valid:

$$\boldsymbol{I}(1^+) = \{1\} \qquad \boldsymbol{I}(2^+) = \{1\}$$
$$\boldsymbol{I}(1^-) = \{1\} \qquad \boldsymbol{I}(2^-) = \{1\}$$

Now supposed that in one of the previous recursion steps for the calculation of the search tree 1^+ has already been applied, i. e. $\boldsymbol{I}(\boldsymbol{J}_k) = \{1^+, \cdots\}$. Hence, for further steps, only the area of the polytope left of j_1 is existent. If the estimation of the remaining region concerning the hyperplane j_2 is applied, one can see that $\{1\}$ is an element of $\boldsymbol{I}(\boldsymbol{J}_k) \cap \boldsymbol{I}(2^+)$ as well as of $\boldsymbol{I}(\boldsymbol{J}_k) \cap \boldsymbol{I}(2^-)$ because $\{1\} \in \boldsymbol{I}(2^+)$ and $\{1\} \in \boldsymbol{I}(2^-)$. In reality, however, "below", i. e. right of j_2, there is no further part of $\mathcal{P}_1$, because this part has already been cut off by the application of 1^+. The calculation with $\boldsymbol{I}_k^- = \boldsymbol{I}(\boldsymbol{J}_k \cup 2^-)$ correctly does not contain $\{1\}$.

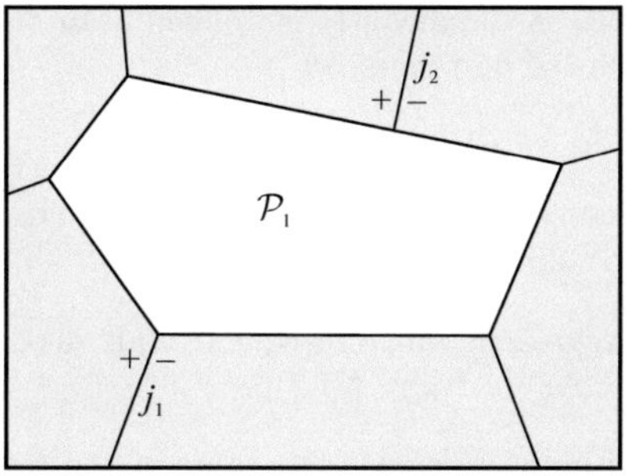

Figure 9.11: Functional principle of the search tree

For the exact index sets, the hyperplane j_k is now determined, for which the value of $\max(|\boldsymbol{F}(\boldsymbol{I}_k^+)|, |\boldsymbol{F}(\boldsymbol{I}_k^-)|)$ becomes minimal. Thus, it can be determined which hyperplane produces the best possible distribution of control laws with the intent to keep the depth of the search tree as small as possible in order to reduce the necessary online computation time. If more than one hyperplane satisfies this optimality constraint the number of remaining polytopes, $\max(|\boldsymbol{I}_k^+|, |\boldsymbol{I}_k^-|)$, is considered as additional criterion, i. e. the hyperplane which reduces the set of polytopes to be evaluated and also the number of remaining control laws, is selected. If there are still several optimum hyperplanes remaining, there is no further evaluation due to reasons of complexity and simply one of the possible hyperplanes is selected.

As it is determined now by which hyperplane the state space is separated in the best way, the information about the tree node N_k can be completed. The optimum hyperplane j_k, determined above, is used for the separation of the tree. Both originating subsequent nodes $N_k^\pm$ are initialized with $\boldsymbol{J}_k^\pm = \boldsymbol{J}_k \cup j^\pm$ and the $\boldsymbol{I}_k^\pm$ determined above. If the number of control laws in the subsequent node is $|\boldsymbol{F}(\boldsymbol{I}^\pm)| > 1$, they are added to the set of nodes which still have to be evaluated. Otherwise the subsequent node is an end node, as for the whole remaining region one single control law is valid. Thus, the number of the corresponding control law is assigned to this end node.

As already mentioned above, these steps are repeated until all nodes are evaluated. Then, the search tree is completely known. The calculation of the binary search tree is mathematically rather complex; due to the fact that the calculation can be done offline, this complexity can be accepted.

Once the calculation of the search tree is done, the online evaluation of the explicit control law is significantly simplified. For this purpose, only the following algorithm has to be processed:

1. Select the root N_1 as current node N_k.
2. Evaluate the describing function $d_j(\mathbf{x})$ from equation (9.27) in which $\mathbf{x}$ is the current system state and $j = j_k$.
3. Branch to the corresponding subsequent node dependent on the sign of $d_j(\mathbf{x})$, i. e. $N_k = N_k^+$ or $N_k = N_k^-$
4. If N_k is an end node, calculate the value of the actuating variable $\mathbf{u}(t)$ from the control law of the end node. Otherwise go back to step 2.

For a practical realization of an MPC controller based on the receding horizon principle, it is important to branch out the tree only at those hyperplanes that separate polytopes from each other whose corresponding control laws differ already in the first step. In this way, an additional simplification of the search tree can be achieved still granting an optimum control method.

9.5.4 Optimal complexity reduction

In many cases the explicit solution of an MPC controller is calculated for a prediction horizon greater than one. Considering especially a discrete controller, a multiplicity of polytopes whose corresponding control laws being basically different, however, not in the first step, do exist. As, in a real control, only the first value of the precalculated sequence of values for the actuating variables is really applied to the plant, it makes sense to combine polytopes whose control laws are the same for the first step in order to reduce the complexity of the structure and therewith also to reduce the necessary computation time. Unfortunately, this combination cannot be realized by simply combining all possible polytopes since it has to be ensured that the resulting structure again consists of convex polytopes (see further in this chapter). In [43], Geyer,

Torrisi and Morari propose a method to combine regions with the same control law as skillfully as possible under consideration of the convexity constraint. If this strategy is applied in dependence of the first step of every control law, a significant simplification of the polytope structure can be expected. Thus one can proceed as follows:

If the calculated piecewise affine controller structure is available as a collection of hyperplanes, a unique marking M for every region can be determined via a sign vector which in principle shows the relation between a certain region and the different hyperplanes. An algorithm doing such a *cell enumeration* is e. g. described in detail by Geyer et al. in [42]; here a detailed description is omitted. The determined markings are the basis for the optimal complexity reduction to be done in the following.

If all boundaries are only defined on state variables and input variables, the hyperplanes and markings are also defined on the state and input variable space. Such a case is called *global hyperplane arrangement*. The markings M of the hyperplane arrangement are then identical to the index sets $\boldsymbol{J}$ of the piecewise affine controller structure that has already been described in chapter 9.5.3 and the single polytopes as well as the single descriptions of the dynamics can be set equal to each other. However, if the boundaries should furthermore be dependent on additional variables which themselves are again dependent on other boundaries, the algorithm described in [42] results in a collection of hyperplane arrangements being sequentially defined into one another. Hence, the set of regions does not cover the whole state space, but it is a polytope subset of it. Such a case is called *local hyperplane arrangement*. It commonly occurs if the system consists of several so-called *discrete hybrid automata* (DHA) which are sequentially dependent on each other, i. e. one DHA defines a group of hyperplanes within a polytope of the group of hyperplanes of the previous PWA description. However, this very complex case is not to be assumed in the application of a drive control with inverter as discussed here. Thus, it can be assumed that it has only to be dealt with a global hyperplane arrangement here.

For a given PWA description, e. g. a controller structure for an explicit MPC controller, now an optimization problem has to be solved. This means that a new set $\{\mathcal{Q}_j\}$, $j = 1 \dots Q$ with the following characteristics has to be created from a given set $\{\mathcal{P}_i\}$, $i = 1 \dots P$ of polytopes which describe the same dynamics:

- The union of the new polytopes is equal to the union of the original ones, i. e. $\bigcup_{i=1}^{P} \mathcal{P}_i = \bigcup_{j=1}^{Q} \mathcal{Q}_j$.

- The new polytopes are mutually disjoint, i. e. the following statement is valid: $\mathcal{Q}_i \neq \mathcal{Q}_j$ for all $i, j \in \{1 \dots Q\}$, $i \neq j$.
- The new polytopes are unions of the old ones, i. e. for each $\mathcal{Q}_j$, $j \in \{1 \dots Q\}$ there is an index set $\boldsymbol{I} \subseteq \{1 \dots P\}$ so that $\mathcal{Q}_j$ is the union of all $\mathcal{P}_i$, $i \in \boldsymbol{I}$.
- The number Q is minimal, i. e. there is no set $\{\mathcal{Q}_j\}$, $j = 1 \dots Q$ with a lesser number of polytopes.

The optimization task is non-trivial because the union of all polytopes with the same description of the dynamics is normally not convex; furthermore, overlapping regions are not allowed.

First, it is to be assumed that besides the PWA description a corresponding global hyperplane arrangement exists together with the associated markings M which can be seen to be given for the existent control structure or be determined easily. With the help of these markings the following characteristics of polytopes can be determined quickly:

Separating hyperplanes: A hyperplane j and two polytopes $\mathcal{P}_1$ and $\mathcal{P}_2$ with the associated markings M_1 and M_2 shall be given. The hyperplane j is a separating hyperplane with regard to the polytopes $\mathcal{P}_1$ and $\mathcal{P}_2$ if M_1 and M_2 differ in the jth element.

Envelope: Two polytopes $\mathcal{P}_1$ and $\mathcal{P}_2$ with the associated markings M_1 and M_2 shall be given, in which $M_1(i) = M_2(i)$ for all $i \in \boldsymbol{J}$ and $M_1(i) \neq M_2(i)$ for all $i \in \{1 \dots L\} \setminus \boldsymbol{J}$. L is the whole number of all hyperplanes. If a new marking M with $M(i) = M_1(i)$ for all $i \in \boldsymbol{J}$ and $M(i) =$ '*' for all $i \in \{1 \dots L\} \setminus \boldsymbol{J}$ is constructed, the envelope $\text{env}(\mathcal{P}_1, \mathcal{P}_2)$ is given by M.

Convexity: Two polytopes $\mathcal{P}_1$ and $\mathcal{P}_2$ with the associated markings M_1 and M_2, $M_1 \neq M_2$ shall be given. The combination $\mathcal{P}_1 \cup \mathcal{P}_2$ is convex if, and only if, the markings M_1 and M_2 differ in exactly one element.

Neighboring polytopes: Two polytopes are called neighboring if they have a common facet.

Connections: A set of polytopes $\{\mathcal{P}_i\}$, $i \in \boldsymbol{I}$ with the associated markings M_i shall be given. The polytopes are connected if another polytope $\mathcal{P}_j$, $j \in \boldsymbol{I}$, $j \neq i$ exists for every polytope $\mathcal{P}_i$, $i \in \boldsymbol{I}$, so that the associated markings M_i and M_j differ in exactly one element.

Having a closer look at the markings of the hyperplane arrangements, it can easily be figured out if two polytopes are connected with each other, if their envelope is convex or similar things. This is used in order to merge those polytopes of the controller structure which contain the same control law with the help of a recursive Branch and Bound algorithm. At first look, this seems to be complicated, however, it is quite simple.

If e. g. the polytopes with the control law x should be optimally combined, the envelope $\mathcal{P}_x = \text{env}(\boldsymbol{M}_x)$ is constructed first, in which $\boldsymbol{M}_x$ describes the set of all markings of these polytopes. As described above, the marking M_x of the envelope can simply be determined by setting the corresponding elements of M_x to the value '*'. If $\mathcal{P}_x$ does not contain any polytopes with other control laws the problem is already solved due to the fact that the combination of all polytopes with the control law x is convex; in this case, the optimum combination simply consists of the combination of all polytopes.

Usually, $\text{env}(\boldsymbol{M}_x)$ will, however, include polytopes with other control laws, i. e. the combination of $\boldsymbol{M}_x$ is not convex. In this case the envelope $\mathcal{P}_x$ has to be reasonably separated. Thereto, $\boldsymbol{J}_x$ shall be the set of indices of all separating hyperplanes with regard to the polytopes within $\mathcal{P}_x$. As the definitions stated above prove, $\boldsymbol{J}_x$ is simply the enumeration of all positions on which the elements of $\boldsymbol{M}_x$ have the value '*' and thus, it can be easily determined. For every hyperplane $i \in \boldsymbol{J}_x$, the envelope can now be separated into two new convex regions which are named $\mathcal{P}_{x,i+}$ and $\mathcal{P}_{x,i-}$ with the associated markings $\boldsymbol{M}_{x,i+}$ and $\boldsymbol{M}_{x,i-}$. The Branch and Bound algorithm now branches at the hyperplane i by calling itself two times; once with $\boldsymbol{M}_{x,i+}$ and once with $\boldsymbol{M}_{x,i-}$ as the polytope structure to be optimized. This step is repeated for all other hyperplanes $i \in \boldsymbol{J}_x$.

In two cases, a node of the tree is an end node: First, if the envelope of the current polytope structure contains no more polytopes with other control laws, i. e. when the combination is convex. The second case applies if the current polytope structure contains no further polytopes with the control law of the polytope structure that should be optimized since in this case no further polytopes are available for merging. Furthermore, the usual techniques for excluding branches from the tree that will recognizably not lead to an optimum, are used.

Since for a practical implementation of an explicit MPC controller, only the first value of the sequence of actuating variables $\mathbf{u}$ is of interest, a reduction of the controller structure is not only done under consideration of the control law itself, but all regions, which will in the next sampling cycle apply the same value of the actuating variable to the plant, are merged. As a two-level inverter with DC link has only eight different switching states, as already explained in chapter 9.2, $\mathbf{u}(t)$ can take one out of only eight values. For controls with tracking, i. e. with any values for the actuating variables, $\Delta\mathbf{u}(t)$ is used as input variable instead of $\mathbf{u}(t)$. Hence, the maximum number of possible values increases to 27 because for the three half bridges applies that Δa is, as well as Δb and Δc, an element of $\{-1, 0, 1\}$. According to this, only 27 different groups of polytopes exist. However, since the convexity constraint has to be regarded, too, it cannot be expected that the controller structure consists of only 27 different polytopes after the optimization process has been performed; the real number of polytopes will be significantly higher.

It has to be mentioned that solutions calculated with the method of Optimal Complexity Reduction have in principle a higher online computational demand than a binary search tree according to chapter 9.5.3, although both algorithms only consider those hyperplanes which lead to different values for the actuating variables in the next sampling cycle, but the search tree further accelerates the online-search for the control law to be applied through the fact that here an easier approach is possible because of the tree-like structure, while, considering the simplified structure obtained by optimum complexity reduction, an exhaustive search over the remaining polytopes has still to be done. However, in experiments it was proven that due to the high memory requirements for very complex solution structures, a direct calculation of the search tree is not possible; however, it is possible to determine the structure with the help of the algorithm described here first and then to determine the binary search tree.

9.5.5 Experimental results

LTI system

For the evaluation of the explicit control, the drive was first considered as a linear time-invariant (LTI) system, i. e. the hybrid structure of the drive was neglected. So the machine model derived in chapter 7 for a multidimensional GPC controller can be used; however, the transformation into the CARIMA model can be omitted, as for the explicit solution, the discrete-time state space model is used. Additionally, the consideration of the cross coupling in form of known disturbances, as described in chapter 7.3, was omitted so that the model used is in fact a discretized form of the simplified model shown in figure 3.2 on page 14. With the help of this model, the computation time for an explicit MPC control can be examined comparatively. Table 9.3 shows the results for the different algorithms. For reasons of comparison, the computational demand for an online calculated MIMO GPC controller is also given.

Type	Structure	Max. calc. time
exhaustive search	225 polytopes	163 µs
Minimum-Time-Controller	9 polytopes, 2 iterations	27 µs
Opt. complexity reduction	60 polytopes	94 µs
Binary search tree	225 nodes, branches: 10	4.7 µs
MIMO-GPC	– analytical solution –	4.1 µs

Table 9.3: Calculation times for MIMO current controllers

As it can easily be seen, model-based predictive control with explicit solution of the optimization task (see chapter 9.5.1) is not feasible for drive control without further modifications since the complete enumeration over the 225 polytopes of the controller structure needs significantly too much time. However, a minimum-time-controller whose execution time is significantly smaller, because only two iterations with altogether nine polytopes have to be evaluated, is feasible (see chapter 9.5.2). Its drawback is, however, that it does not provide an optimum control considering the cost function. The use of optimal complexity reduction (see chapter 9.5.4) also leads to an acceleration of the online search because of the simplification of the controller structure from 225 to only 60 remaining polytopes; due to the exhaustive search over the remaining 60 polytopes, which is still necessary, the computation time is still too high. When a binary search tree (see chapter 9.5.3) is used, the online calculation time can be reduced by a factor of nearly 35 compared to the standard method which makes an implementation feasible. An interesting fact is that the search tree also contains 225 elements, i. e. the same amount of elements as the original polytope structure. However, by arranging the optimum solution as a tree-like structure, in the worst case only ten decisions have to be made, significantly less than the 225 for the exhaustive search. Furthermore, every decision in the search tree is always coupled to the evaluation of only *one* hyperplane equation; for the polytope search, *all* hyperplanes of a polytope have to be evaluated, if possible, in order to determine if a given point lies within this polytope or not.

In summary it can be stated that explicit MPC for an electrical drive modeled as LTI system is feasible considering the computation time. However, in comparison to the analytical solution, no reduction of the computation time could be determined. This is why, in this case, the analytical GPC method should be preferred.

Hybrid system

As the measured computation times given in table 9.3 generally prove the feasibility of explicit MPC also for electrical drive technology, now a *direct* model-based predictive control scheme with explicit solution is presented. Therefore, at first, the simplified machine model explained in chapter 9.3.1 has to be discretized; the procedure is the same as described in chapter 7. For the discrete-time simplified machine model in stator coordinates according to the equations (9.10) and (9.11), the following system matrices result:

$$\mathbf{A} = \begin{bmatrix} 1-\dfrac{T_0}{\tau_\sigma{}'} & 0 \\ 0 & 1-\dfrac{T_0}{\tau_\sigma{}'} \end{bmatrix} = \begin{bmatrix} 0.9873 & 0 \\ 0 & 0.9873 \end{bmatrix} \tag{9.28}$$

$$\begin{aligned} \mathbf{B} &= \begin{bmatrix} \dfrac{2\,T_0}{\sqrt{3}\,r_\sigma \tau_\sigma{}'} & -\dfrac{T_0}{\sqrt{3}\,r_\sigma \tau_\sigma{}'} & -\dfrac{T_0}{\sqrt{3}\,r_\sigma \tau_\sigma{}'} \\ 0 & \dfrac{T_0}{r_\sigma \tau_\sigma{}'} & -\dfrac{T_0}{r_\sigma \tau_\sigma{}'} \end{bmatrix} \\ &= \begin{bmatrix} 0.1713 & -0.08566 & -0.08566 \\ 0 & 0.1484 & -0.1484 \end{bmatrix} \end{aligned} \tag{9.29}$$

$$\mathbf{C} = \mathbf{I} \tag{9.30}$$

$$\mathbf{D} = \mathbf{0} \tag{9.31}$$

If the usual algorithms for the calculation of an explicit solution are applied to these matrices, a controller structure results, which neither considers the delay of one sampling cycle caused by the digital control nor is feasible for controlling to another reference value than $\mathbf{u}(t) = [0\ 0]^T$. Thus, the matrices of the equations (9.28)–(9.31) must be transformed to the corresponding matrices under consideration of delay and tracking according to the equations (9.22) and (9.23). This gives for matrix $\mathbf{A}_{t,d}$

$$\mathbf{A}_{t,d} = \begin{bmatrix} 1-\dfrac{T_0}{\tau_\sigma{}'} & 0 & \dfrac{2\,T_0}{\sqrt{3}\,r_\sigma \tau_\sigma{}'} & -\dfrac{T_0}{\sqrt{3}\,r_\sigma \tau_\sigma{}'} & -\dfrac{T_0}{\sqrt{3}\,r_\sigma \tau_\sigma{}'} & 0 & 0 \\ 0 & 1-\dfrac{T_0}{\tau_\sigma{}'} & 0 & \dfrac{T_0}{r_\sigma \tau_\sigma{}'} & -\dfrac{T_0}{r_\sigma \tau_\sigma{}'} & 0 & 0 \\ 0 & 0 & 1 & 0 & 0 & 0 & 0 \\ 0 & 0 & 0 & 1 & 0 & 0 & 0 \\ 0 & 0 & 0 & 0 & 1 & 0 & 0 \\ 0 & 0 & 0 & 0 & 0 & 1 & 0 \\ 0 & 0 & 0 & 0 & 0 & 0 & 1 \end{bmatrix}$$

or, represented in numerical values

$$\mathbf{A}_{t,d} = \begin{bmatrix} 0.9873 & 0 & 0.1713 & -0.08566 & -0.08566 & 0 & 0 \\ 0 & 0.9873 & 0 & 0.1484 & -0.1484 & 0 & 0 \\ 0 & 0 & 1 & 0 & 0 & 0 & 0 \\ 0 & 0 & 0 & 1 & 0 & 0 & 0 \\ 0 & 0 & 0 & 0 & 1 & 0 & 0 \\ 0 & 0 & 0 & 0 & 0 & 1 & 0 \\ 0 & 0 & 0 & 0 & 0 & 0 & 1 \end{bmatrix}$$

and for the remaining system matrices

$$\mathbf{B}_{t,d} = \begin{bmatrix} 0 & 0 & 0 \\ 0 & 0 & 0 \\ 1 & 0 & 0 \\ 0 & 1 & 0 \\ 0 & 0 & 1 \\ 0 & 0 & 0 \\ 0 & 0 & 0 \end{bmatrix}$$

$$\mathbf{C}_{t,d} = \begin{bmatrix} 1 & 0 & 0 & 0 & 0 & 0 & 0 \\ 0 & 1 & 0 & 0 & 0 & 0 & 0 \\ 0 & 0 & 1 & 0 & 0 & 0 & 0 \\ 0 & 0 & 0 & 1 & 0 & 0 & 0 \\ 0 & 0 & 0 & 0 & 1 & 0 & 0 \\ 0 & 0 & 0 & 0 & 0 & 1 & 0 \\ 0 & 0 & 0 & 0 & 0 & 0 & 1 \end{bmatrix}$$

$$\mathbf{D}_{t,d} = \begin{bmatrix} 0 & 0 & 0 \\ 0 & 0 & 0 \\ 0 & 0 & 0 \\ 0 & 0 & 0 \\ 0 & 0 & 0 \\ 0 & 0 & 0 \\ 0 & 0 & 0 \end{bmatrix}$$

As not the actuating variable itself should be evaluated, but only the change of the value of the actuating variable, which corresponds to the switching effort of the inverter, the weighting matrices have to be defined in the following way

(λ is the weighting factor for the change of the value of the actuating variables):

$$\mathbf{Q}_{t,d} = \begin{bmatrix} 1 & 0 & 0 & 0 & 0 & -1 & 0 \\ 0 & 1 & 0 & 0 & 0 & 0 & -1 \\ 0 & 0 & 0 & 0 & 0 & 0 & 0 \\ 0 & 0 & 0 & 0 & 0 & 0 & 0 \\ 0 & 0 & 0 & 0 & 0 & 0 & 0 \\ -1 & 0 & 0 & 0 & 0 & 1 & 0 \\ 0 & -1 & 0 & 0 & 0 & 0 & 1 \end{bmatrix}$$

$$\mathbf{R}_{t,d} = \begin{bmatrix} \lambda & 0 & 0 \\ 0 & \lambda & 0 \\ 0 & 0 & \lambda \end{bmatrix}$$

For the model described above, controller structures with around 7500–8500 polytopes result; the exact number is varying a little bit depending on the value of λ. However, commonly it can be stated that a higher weighting of $\Delta\mathbf{u}$ leads to simpler controller structures. Nevertheless, an explicit controller of this complexity is not feasible due to reasons of memory demands and especially due to the needed computation time. The offline computation of a search tree is also not possible since the memory requirements for these tasks are not available on conventional PCs. A workaround is to reduce the complexity beforehand to around 1800–2200 polytopes by merging regions with the same $\Delta\mathbf{u}$; after that, the calculation of a binary search tree is possible. Finally, a tree-like structure results with around 2100–2300 nodes and a maximum tree depth of 18 which needs less than 8 µs of online computation time on the used computer system. If this computation time is compared with the times given in table 9.2, one can see that model-based predictive control with direct control of the inverter is feasible with the help of an explicit solution, in contrast to the online optimization methods discussed there.

The behavior of such a control was analyzed using the example of current control of an electrical drive; the experimental results can be seen in figure 9.12. Also here, as in chapter 9.4.3, measurement results are given in a stationary coordinate frame; here again, only a step change of the reference value $i_{s\alpha}$ was done, in order to get some information about the coupling between $i_{s\alpha}$ and $i_{s\beta}$. For a comparison with a conventional control scheme, the same measurement was repeated with PI current controllers for $i_{s\alpha}$ and $i_{s\beta}$ (figure 9.12(c)), whereas here, of course, a modulator (PWM) has to be inserted between controller and inverter. In order to get, despite of this fact, reasonably comparable results, the switching frequency of the inverter was, for these measurements, reduced to 1630 Hz, for direct control, however, a sampling frequency of 9770 Hz was used.[3]

As it can be seen in figure 9.12, the direct explicit MPC has a behavior similar to the direct GPC in figure 9.8. The switching frequency and thus, the value of the control deviation can be influenced via the weighting factor λ, while the dynamics of the complete system are not affected by this. The new reference value will always be reached in the shortest possible time, i. e. the dynamics of the controlled plant are only limited by the physical limits of the system itself. Weighting the switching events less than $\lambda = 0.1$ does not make sense because, for the selected sampling time of around 100 µs, the minimum of the current disturbances is already reached, as it can be seen in figure 9.12(a).

An influence of the cross coupling between the stator current $i_{s\alpha}$ and $i_{s\beta}$ cannot be figured out because it is significantly smaller than the amplitude of the current harmonics. Thus, an implementation of this cross coupling in the plant model does not make sense since it would result only in a higher, but unnecessary complexity of the explicit controller structure.

Figure 9.12(c) shows the behavior of the system when using a linear PI controller for the same control task and the same sequences of reference values as used for direct control in figure 9.12(a) and figure 9.12(b). As it is to be expected, the current distortions are much smaller when a linear controller is used because using an inverter, controlled via a modulator, the switching times of the semiconductors are variable within a sampling cycle, while direct controllers can only switch at the beginning of every sampling cycle. However, a drawback of a PI controlled drive is that the dynamics of the complete system are coupled to the switching frequency, because the switching frequency is dependent on the sampling frequency and for a PI controller it does not make sense to use a higher sampling frequency than twice the switching frequency. The reason for this is the fact that, for parameterizing a PI controller, the dead time caused by the digital control has to be considered, which leads to a more sluggish control. However, this dead time is directly related to the selected sampling rate. If a low switching frequency is desired, e. g. for a high power converter, the dynamics of the complete system are degraded if current control is done by PI controllers because the low switching frequency also causes a low sampling rate. However, when using an explicit DMPC controller, it is possible to use a comparably high sampling rate and to ensure, despite of this, a low average switching frequency by using a considerably high weighting of the

[3] Since six switching instances per period of the carrier signal take place when using modulator-based inverter control, the switching frequency for the measurements with a PI controller was set to 1/6th of the sampling frequency of the direct controller.

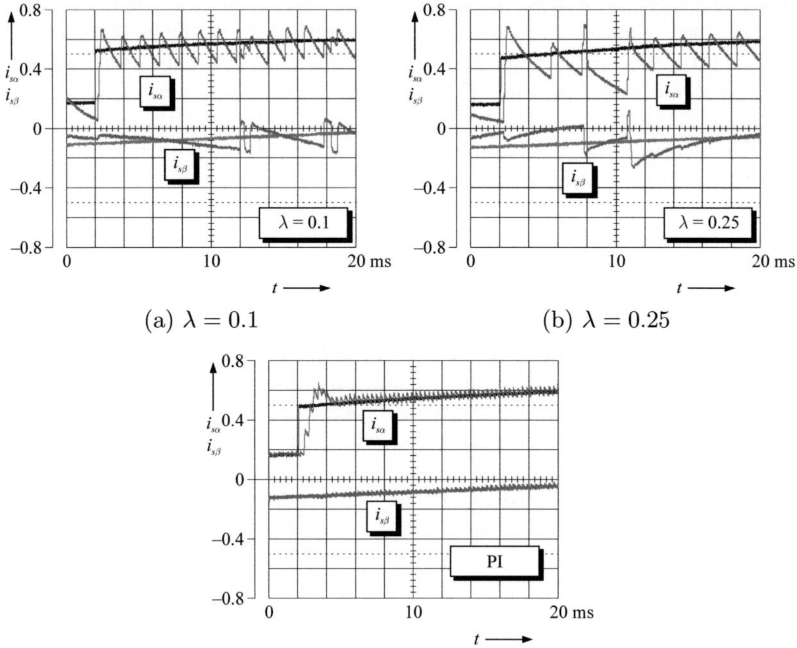

(a) $\lambda = 0.1$

(b) $\lambda = 0.25$

(c) PI controller

Figure 9.12: Current control with explicit DMPC and PI controller

switching events in the cost function. Furthermore, in model-based predictive controllers, the dead time caused by digital controllers can be considered in the controller design. Thus, this dead time has a significantly lesser influence on the dynamics of the controlled system.

10 Related control structures

In literature, one can quite often find control methods being partly denoted as model-based predictive controllers or which are said to be equal with some predictive control strategies. Here, especially *Internal Model Control* and *Linear Quadratic Control* have to be mentioned. In fact, both controllers have some similarities compared to MPC that cannot be overlooked. In the following sections, these similarities, but also the differences of these methods compared to a real model-based predictive control strategy, will be presented.

10.1 Internal Model Control

Internal Model Control (IMC) is actually no *control method*, but a special *design principle* for a control. However, it is, probably because of its name, assigned to MPC controllers by some authors, e. g. García/Prett/Morari [40]. In a very detailed description of the method by Lunze [84], it is also introduced under the name "model-based control"; however, in this context it is not assumed to be a predictive or precalculating control. In order to show that the basic ideas of IMC are, despite some similarities, significantly different from the ones on which MPC is based, they will shortly be outlined here. For further information, the interested reader is referred to the book of Lunze [84, chapter 12.2]; a detailed comparison between MPC and IMC can be found in Dittmar/Pfeiffer [34, chapter 2.4.2]. By Harnefors/Nee [48], the application of IMC for current control of an induction machine is described in detail.

The basic principle of IMC controller design is to integrate an inner model of the plant to be controlled into the controller itself. Thus, it is contrary to conventional controls which use a plant model only for parameterizing the controller; the actual control law itself does, however, not contain a model. Figure 10.1 shows the typical structure of an IMC controller.

The plant model is connected in parallel to the plant which leads to the fact that not the controlled variable $\mathbf{y}(t)$, but the difference between the real and the output vector calculated via the model, $\mathbf{y}(t)-\hat{\mathbf{y}}(t)$, is fed back to the controller. A value of this feedback unequal to zero can be caused by two different things:

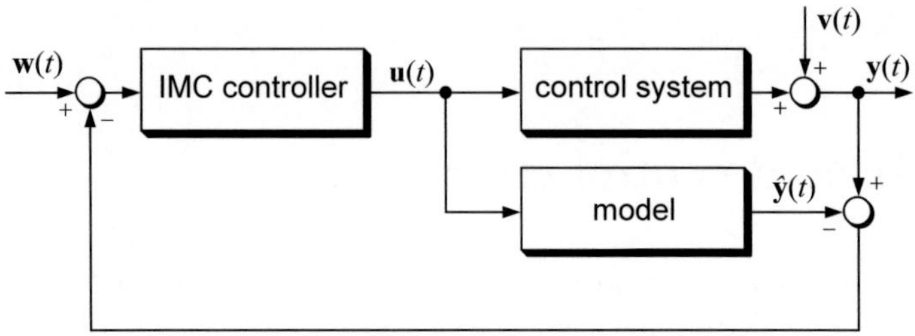

Figure 10.1: IMC structure

1. A disturbance $\mathbf{v}(t)$ affects the plant.

2. A model error is present.

Assuming that the model is not affected with an error and that there is no disturbance $\mathbf{v}(t)$ existing, the feedback is also equal to zero. So the IMC controller is, in this case, a simple open-loop control and offers a holistic approach for the direct selection of the controller structure by deducing it from the desired behavior of the entire system.

If a disturbance $\mathbf{v}(t)$ occurs, the difference $\mathbf{y}(t)-\hat{\mathbf{y}}(t)$ will be unequal to zero. The controller reacts to this signal by correspondingly changing the controller output $\mathbf{u}(t)$. This ability to react to disturbances is a difference of an IMC controller compared to a mere feedforward control. No stability problems will occur; if the plant and the IMC controller are stable, the closed control loop will also be stable.

For a practical realization of an IMC controller, the IMC controller structure (figure 10.1) is transformed into the conventional structure according to figure 10.2; again a feedback of the controlled variable $\mathbf{y}(t)$ results. Thereby, the dashed box corresponds to a conventional controller $F_R(s)$. If the plant model is known and the IMC controller is designed according to the desired behavior (see above), $F_R(s)$ can easily be determined.

In reality, one cannot assume that the plant model is in all cases equal to the real plant. The possibilities to consider this in an IMC controller are described by Harnefors/Nee [48] and Zafiriou/Morari [123] and thus, they are not further discussed here. Thereby, also an adaptive plant model is possible.

Compared to conventional control with linear controllers, IMC has some advantages; among other things, the controller design is significantly easier. If,

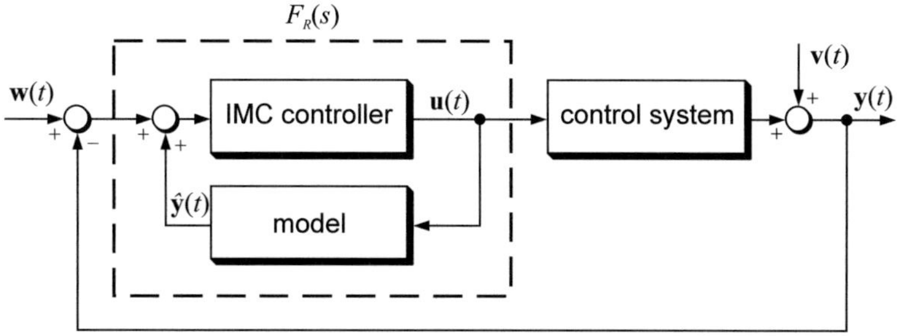

Figure 10.2: IMC structure transformed to the classical controller structure

e. g., the IMC structure is translated into a conventional controller structure according to figure 10.2 and a PI controller is used, the proportional and integral gains of the controller are directly expressed in terms of the machine parameters and the desired controller bandwidth. Furthermore, the stability of the complete system can always be granted if the system to be controlled is open-loop stable and if the IMC controller is stable, too.

Although Internal Model Control apparently seems to be an attractive design principle, in reality some problems occur which lead to the fact that an IMC controller can practically not be realized. Besides stability of the actual IMC controller itself, causality of the control law has to be guaranteed, too. Dittmar/Pfeiffer [34, chapter 2.4.2] show that even for a plant only consisting of a simple PT_1-block with dead time no realizable IMC controller can be found since the inversion of the dead time leads to the fact that, for the determination of the current value of the actuating variable, future values of the control deviation are needed. If an IMC controller should be used despite this fact, the model of the plant to be controlled has to be simplified in such a way that the design of a practically realizable controller according to the basics of IMC is possible, which consequently leads to a non-optimum control in this case.

10.2 Linear Quadratic Control

Sometimes, it is told that Generalized Predictive Control, a subset of model-based predictive controllers presented in chapter 6 and 8 for uni- and multi-dimensional plants, is identical to *Linear Quadratic Control*. However, this is *definitely not* true. Both methods are similar and in some special cases, one can obtain the same results with both methods, although these are clearly two different control strategies.

10.2.1 Functional principle of LQR

Before comparing the methods GPC and LQR, it is necessary to explain the basic principles of Linear Quadratic Control. For reasons of simplicity, only the basic design principle of an LQ control for a deterministic time-invariant system is described. The interested reader shall be refered to further literature. A detailed description of the LQR principle can be found e. g. in the publication of Isermann [58, chapter 8.1] or Kanjilal [62, chapter 13]; a good explanation for the case of a continuous-time LQ controller has been written by Lunze [85, chapter 7].

Linear Quadratic Control is based on a state space model of the plant according to the equations (9.10) and (9.11) on page 143 as well as the corresponding cost function (9.12)[1]. Since for an LQR control it is always assumed that the system variables are not constrained, the following statements are valid: $\mathbf{x} \in \mathrm{IR}^n$, $\mathbf{y} \in \mathrm{IR}^p$ and $\mathbf{u} \in \mathrm{IR}^m$; furthermore, for LQR, only one single horizon exists. Thus, $N_1 = 0$ and $N_2 = N_p = N_u = N$. In order to avoid mixing up with the weighting matrix $\mathbf{P}$ that will be introduced in the following, the weighing of the final value of the state vector is denoted with $\mathbf{S}$ here. Thus, the cost function can be written as:

$$\begin{aligned} J(N) = {} & \mathbf{x}(t+N)^T \, \mathbf{S} \, \mathbf{x}(t+N) \\ & + \sum_{j=1}^{N} \Big(\mathbf{x}(t+j-1)^T \, \mathbf{Q} \, \mathbf{x}(t+j-1) \\ & \qquad + \mathbf{u}(t+j-1)^T \, \mathbf{R} \, \mathbf{u}(t+j-1) \Big) \end{aligned} \tag{10.1}$$

Now, the intention is to obtain an optimum control law with which according to the equation

$$\mathbf{u}(t) = -\mathbf{K}(t)\mathbf{x}(t) \tag{10.2}$$

optimum future values for the actuating variables can be calculated. Thereby, $\mathbf{K}$ denotes the so called *controller matrix* of the LQ controller. Now, such a controller matrix $\mathbf{K}$ that minimizes the cost function (10.1) has to be found.

According to the optimality constraint by Bellman [9], the end of an optimum sequence is also an optimum. Consequently, if the last value of an optimum

[1] In order to keep the mathematical representation of LQR consistent to the equations in chapter 9, the index d is not used despite of the discrete-time form.

sequence is known, the other elements can be calculated backwards. Thus, for the solution of the optimization problem, only the last step of the cost function (10.1) is considered:

$$J_N = \mathbf{x}(t+N)^T \, \mathbf{P}_N \, \mathbf{x}(t+N) \qquad\qquad \mathbf{P}_N = \mathbf{S}$$

Thereby, $\mathbf{P}$ is the new weighting matrix. For the above optimization task, an optimum value for $\mathbf{u}(t+N-1)$ is determined by setting the derivation of the cost function J_N equal to zero:

$$\begin{aligned}\mathbf{u}(t+N-1) &= -\left(\mathbf{B}^T\mathbf{S}\mathbf{B}+\mathbf{R}\right)^{-1}\mathbf{B}^T\mathbf{S}\mathbf{A}\,\mathbf{x}(t+N-1)\\ &= -\mathbf{K}_{N-1}\,\mathbf{x}(t+N-1)\end{aligned}$$

Now, the remaining equations for $\mathbf{u}(t+j)$, $j = (N-2)\ldots 0$ and with these, also the equations for the remaining elements of $\mathbf{K}$ and $\mathbf{P}$ can be determined recursively.

Commonly, one can be sure that the controller matrix is not constant. However, if an infinite optimization horizon is used, a time-invariant control law can be obtained:

$$\mathbf{u}(t) = -\mathbf{K}\,\mathbf{x}(t) \tag{10.3}$$

In this case, the controller matrix results in:

$$\mathbf{K} = \left(\mathbf{R}+\mathbf{B}^T\mathbf{P}\mathbf{B}\right)^{-1}\mathbf{B}^T\mathbf{P}\mathbf{A} \tag{10.4}$$

with $\mathbf{P}$ as a solution of the matrix Riccati equation

$$\mathbf{P} = \mathbf{Q}+\mathbf{A}^T\mathbf{P}\left(\mathbf{I}-\mathbf{B}\left(\mathbf{R}+\mathbf{B}^T\mathbf{P}\mathbf{B}\right)^{-1}\mathbf{B}^T\mathbf{P}\right)\mathbf{A} \tag{10.5}$$

The solution of equation (10.5) is done via the recursive calculation stated above. Afterwards, the controller matrix $\mathbf{K}$ for the advance calculation of optimum future sequences of values for the actuating variables is known. Figure 10.3 shows the structure of the complete LQ control. For reasons of simplicity, the feedforward matrix is considered to be $\mathbf{D} = \mathbf{0}$, an assumption that can be made for most cases in drive control. Furthermore, it has to be considered that for the LQ controller shown here, the state variables are controlled to be zero, i. e. a reference value different from the zero vector is not possible. For changing reference values, the control deviation $\mathbf{y}-\mathbf{w}$ or $\mathbf{x}-\mathbf{w}$, respectively, has to be minimized instead of the system state $\mathbf{x}$. The corresponding procedure is analog to the one shown in chapter 9.5 on page 148 et seqq.

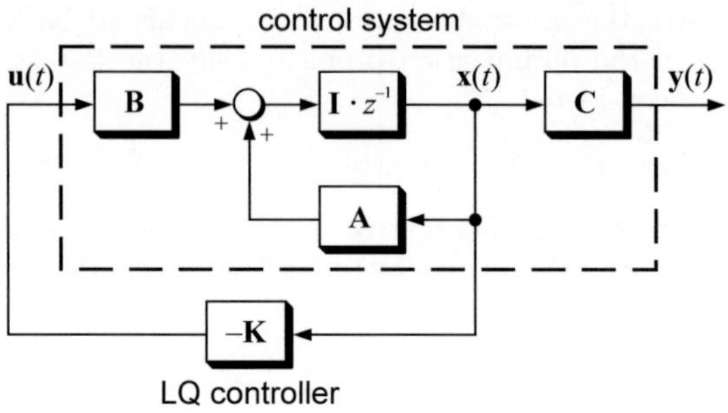

Figure 10.3: Structure of LQ control ($\mathbf{w} = 0$, see text)

10.2.2 GPC and LQR

If Linear Quadratic Control is compared to GPC by all means, some similarities can be found (see also [28, 40, 62])):

- Both methods use a linear model of the plant and GPC as well as LQR minimize a scalar quadratic cost function.
- The optimization problem is solved analytically in both cases, i. e. by differentiating the cost function with regard to the actuating variable and then by searching the zero point.
- Under certain circumstances LQR as well as LRPC deliver identical results.
- LQ controllers with finite horizon as well as GPC controllers minimize the same cost function and the same values for the actuating variables are obtained. Thus, they have the same stability characteristics.

Besides these similarities, there are also significant differences [28, 34, 40, 62]:

- Whereas for GPC a transfer function-based model is used (see chapter 6.1.1 on page 39), the calculations for LQR are based on a state space model.
- Like all strategies based on LRPC, GPC uses the current time t as reference point and from this point, the optimum vector for the actuating

variables $\mathbf{u}(t)$ is calculated based on the advance calculation of the control variables $\mathbf{y}(t+j)$ over a finite horizon into the future. However, for LQR, which is based on a state space representation, the selected reference time point is at the end of the horizon, i. e. at $t+N$, and from that point the values of the actuating variable $\mathbf{u}(t)$ are determined via backward recursions. In this case, the horizon N can, in contrast to GPC, be finite as well as infinite.

- For LQ control, the optimum feedback strategy is calculated in the design process, i. e. offline. For GPC, the optimum sequence of values for the actuating variables is newly determined for every sampling cycle.
- GPC determines an optimum sequence of values for the actuating variables with a finite horizon, in which the cost horizon N_u can in fact be smaller than the prediction horizon N_p. In contrast to that, LQR determines an optimum feedback matrix with a mostly infinite time horizon. If a finite horizon is chosen, cost and prediction horizon are always equal.
- MPC strategies without consideration of constraints—as it is often the case for GPC controllers—result in a time-invariant control law. In contrast to that, LQR commonly leads to a non-constant controller matrix $\mathbf{K}(t)$. A time-invariant controller, which is desired for most applications, can also be determined with LQR, as long as the weighting matrices $\mathbf{S}$, $\mathbf{Q}$ and $\mathbf{R}$ are chosen to be constant over the time and an infinite optimization horizon is selected.
- LQR solves the optimization problem via a Riccati equation. This requires the actuating variable being neither constrained in the slope nor in the amplitude. This assumption will in reality never be fulfilled. In contrast to that, GPC allows an optimization under consideration of such constraints.

As the differences show, GPC is based on different basic ideas than LQR, although some similarities can clearly be seen. Although LQR with finite time horizon and GPC deliver the same results, GPC should be preferred. On the one hand, LQR does in its approach not allow to consider possible system constraints; on the other hand, a GPC method is superior to an LQR strategy considering the practical aspect. LQ controls calculate future values for the actuating variables directly from the state vector. If not all state variables are directly measurable, a state observer is needed. Thus, additional online matrix operations are necessary, leading to the fact that LQR will mathematically be very complex.

11 Summary and future prospects

In the presented work, the fundamental applicability of model-based predictive control in drive technology with the technical means being available today could be proven. Computation times that do even satisfy the high performance needs of a current controller were achieved; the computation times are consistently lower than 10 µs. However, some constraints, which do still limit the applicability of MPC, have to be considered.

The GPC principle, which is based on an analytical solution of the optimization problem, is applicable to linear systems without any problems. Even a speed controller with a prediction horizon of $N_p = 200$ needs a computation time of less than 9 µs. An advantage of the GPC method is that because of the transfer function-based model structure, filtering of the measured values is possible without delays. Furthermore, constant disturbances are considered by the controller, since former values of the actuating and of the control variables are also used for the optimization.

For a mere current controller (see chapter 6.3.3 on page 60 et seqq.), the advantages of a GPC controller are only noticeable considering small-signal behavior. Equal results can also be achieved by using conventional techniques, i. e. with a PI controller together with a feed-forward control. Thus the additional effort for a GPC current controller is not really rewarding. The same statement can be given for a MIMO current controller for i_{sd} and i_{sq} used in field-oriented control (see chapter 8.4 on page 115 et seqq.). Indeed, the consideration of the cross coupling between the flux- and torque-producing components of the stator current $\boldsymbol{i}_s$ shows an improvement of the controller dynamics; however, similar results can also be achieved—with less effort—with an adequate extension of the PI control.

However, GPC speed control (see chapter 6.3.3 on page 63 et seqq.) shows significant advantages compared to linear control, since the speed information derived from an incremental encoder signal has to be filtered quite intensively when using conventional control methods. Here the possibility to integrate a delay-free filtering into a CARIMA model structure according to chapter 6.2 on page 48 et seqq. becomes positively noticeable. Thus GPC for use in a speed control application is profitable.

However, a drawback of a GPC-based method is that constraints can hardly be implemented because of the analytical solution. Considering a MIMO control for current and speed, it is necessary to limit the machine current. However, as in such a case the current value is not known outside the controller, the limitation has to be done within the controller structure leading to some problems. For this reason, a practical realization of such a control scheme was given up. Instead, the idea of a direct machine control was pursued.

A direct model-based predictive control has many advantages that cannot be achieved with any other control method. No other method offers the possibility to evaluate the switching effort in such a simple way. Furthermore, switching states that are not allowed, can a priori be excluded from the optimization (see chapter 9.2.1 on page 127 et seq.). Since, considering a GPC method, future values of the actuating variables within the prediction horizon can be forwarded to the controller, control can be done in a stationary coordinate frame without a contouring error (see chapter 9.4.3 on page 138 et seqq.). Unfortunately, a GPC based direct control method can practically not be realized, as the computation time necessary for the evaluation of the optimization problem cannot be reduced to an acceptable value, even if intelligent search strategies are applied (see chapter 9.4.4 on page 141 et seq.).

A way out of this dilemma is the explicit solution of the optimization problem. Unfortunately, this can, until now, only be done for MPC controllers based on state space representation; thus, an explicit GPC control with the methods currently available can at least presently not be realized. Hence, the GPC specific advantages like internal filtering and consideration of past values cannot be used. However, even the explicit solution can only be practically applied when additional tools are used, since the computation times for an online search for the active control law even for small controller structures quickly exceed the acceptable value for drive control. Simplifications and transformations of the explicit controller structure, as they are described in the chapters 9.5.3 and 9.5.4, allow indeed the execution of even complex controllers in very short times.

For current control with a linear model, i. e. with modulator, the explicit MPC solution is inferior to a GPC controller, since the computational effort is nearly the same. GPC, however, offers further advantages due to the reasons stated above. Thus, an explicit MPC controller with linear model cannot be recommended for electric drive technology.

A direct control with a GPC-based method and without a modulator is not feasible because of the needed computation time. However, a transformation of the optimization problem into a polytope structure with affine control laws allows the execution of MPC control of an electric drive under consideration of the value discrete nature of the actuator. An application as current controller shows the superiority of DMPC compared to a linear controller in dynamic mode (see chapter 9.5.5 at page 162 et seqq.). However, some drawbacks do exist:

- System models based on the transfer function are similar to the IIR filters known from communications technology. Hence, a filtering of measured signals can be easily implemented. State space models, on the other hand, offer no possibility to perform a delay-free filtering without any additional effort. If this is needed, a Kalman filter or similar tools have to be included in the signal path. For realizing a current control of an electric drive via DMPC, this is indeed not necessary, since the current distortions are quasi included in the model by regarding the hybrid characteristics of the system. Hence, the disadvantage of the missing delay-free filtering has no impact on the controller performance in this case. However, if a speed control is to be included in the DMPC controller, problems are to be expected.

- For the experimental investigations, a very simple machine model was used, but even this model resulted in controller structures with up to 10,000 polytopes. More complex plant models, which e. g. consider the cross coupling or which allow MIMO control of current and speed, will quickly lead to even more complex structures which are not feasible anymore with the hard- and software currently available. Similar statements can be made for larger control horizons.

- In explicit MPC methods, a consideration of future reference value trajectories is not possible, since in this case for every future sampling cycle the dimension of the controller structure would be enlarged by one, which would again lead to structures that are not feasible anymore. Since current control is done in stator coordinates contouring errors cannot be avoided.

- With the use of the implicit single-step optimization method by means of an explicit solution, the advantage that dynamic constraints for the values of the actuating variables can be considered, as they can e. g. be caused by a bootstrap driver circuit of the inverter, does not apply. Only the approach known from conventional direct control methods remains, i. e. to change the switching state determined by the control afterwards if it is necessary.

- MPC controllers based on state space models cannot consider past input and output variables of the plant and thus they fail in controlling constant disturbances to zero. Unfortunately, for current control of electrical drives the back EMF of the machine is such a disturbance. In order to achieve offset-free control additional effort is necessary.

In summary, it can be stated that model-based predictive controllers can be applied in drive technology. However, the application of MPC in fast switching inverters does not seem to be expedient for the reasons stated above. Direct control of electric drives without intermediary modulator with explicit MPC seems to be much more promising here. However, the current state of research does not allow to make use of the fundamental advantages of direct LRPC. Indeed, the available processing power will surely increase. Furthermore, the worldwide research activities in the area of hybrid systems and explicit MPC control are far from being concluded. Thus further improvements of the offline calculations which lead to smaller controller structures can be expected. Both together will in some time make DMPC feasible for more complex calculations, too. Thus the following intentions seem to be reachable:

- The machine model can be improved in the same degree as in which more complex structures will become feasible. A MIMO control including speed or even position control will then be realizable.

- As for MPC, the cost function can be chosen freely, it is also possible to weight the distortion factor d besides the current error and the switching effort. With these demands *online optimized pulse patterns* can be obtained. The conventional method via offline optimized pulse patterns for different stationary operating points and the exhausting compensation of the modulation error in dynamic mode [15] would then not be necessary anymore.

Finally, nonlinear predictive controllers shall be pointed out. If a DMPC controller is used, the dead time and the value-discrete character of the actuating variable (quantization) caused by the digital control are implicitly considered. Moreover, an electric drive has indeed further nonlinearities, like e. g. the cross coupling of both stator current components or saturation effects in the machine, which are not considered in the simple machine model used here. However, by taking the approach to describe the plant as a piecewise affine system, these nonlinearities could be approximated. Assuming the fact that in the near future also for complex systems a feasible controller structure can be determined, it should be possible that further improvements of DMPC control can be made.

Bibliography

[1] M. Aaltonen, P. Tiitinen, J. Lalu, S. Heikkilä, Direkte Drehmomentregelung von Drehstromantrieben (Direct Torque Control of Three-Phase Drives), *ABB Technik*, No. 3, 1995, pp. 19–24

[2] A. Ackva, H. Reinold, R. Olesinski, A Simple and Self-Adapting High-Performance Current Control Scheme for Three Phase Voltage Source Inverters, *23rd IEEE Power Electronics Specialists Conference pesc'92*, Vol. 1, pp. 435–442, Toledo, 1992

[3] P. Albertos, R. Ortega, On Generalized Predictive Control: Two Alternative Formulations, *Automatica*, Vol. 25, No. 5, 1989, pp. 753–755

[4] V. Ambrožič, R. Fišer, D. Nedeljković, Direct Current Control—A New Current Regulation Principle, *IEEE Transactions on Power Electronics*, Vol. 18, No. 1, 2003, pp. 495–503

[5] V. Ambrožič, G. S. Buja, R. Menis, Band-Constrained Technique for Direct Torque Control of Induction Motor, *IEEE Transactions on Industrial Electronics*, Vol. 51, No. 4, 2004, pp. 776–784

[6] G. Amler, A PWM Current-Source Inverter for high Quality Drives, *EPE Journal*, Vol. 1, No. 1, 1991, pp. 21–32

[7] G. M. Asher, M. Sumner, F. Cupertino, A. Lattanzi, Direct Flux Control of Induction Motor Drives, *European Conference on Power Electronics and Applications EPE 2001*, Graz, 2001

[8] C. Attaianese, V. Nardi, A. Perfetto, G. Tomasso, Vectorial Torque Control: A Novel Approach to Torque and Flux Control of Induction Motor Drives, *IEEE Transactions on Industry Applications*, Vol. 35, No. 6, 1999, pp. 1399–1405

[9] R. Bellman, *Dynamische Programmierung und selbstanpassende Regelprozesse (Dynamic Programming and Self-Adapting Control Processes)*, Munich, Vienna: R. Oldenbourg Verlag, 1967

[10] A. Bemporad, M. Morari, Control of systems integrating logic, dynamics, and constraints, *Automatica*, Vol. 35, No. 3, 1999, pp. 407–427

[11] A. Bemporad, M. Morari, V. Dua, E. N. Pistikopoulos, The Explicit Solution of Model Predictive Control via Multiparametric Quadratic Programming, *American Control Conference ACC2000*, pp. 872–876, Chicago, 2000

[12] A. Bemporad, F. Borrelli, M. Morari, Piecewise Linear Optimal Controllers for Hybrid Systems, *American Control Conference ACC2000*, pp. 1190–1194, Chicago, 2000

[13] A. Bemporad, M. Morari, V. Dua, E. N. Pistikopoulos, The explicit linear quadratic regulator for constrained systems, *Automatica*, Vol. 38, No. 1, 2002, pp. 3–20

[14] R. E. Betz, B. J. Cook, S. J. Henriksen, A Digital Current Controller for Three Phase Voltage Source Inverters, *IEEE IAS Annual Meeting*, Vol. 1, pp. 722-729, New Orleans, 1997

[15] B. Beyer, *Schnelle Stromregelung für Hochleistungsantriebe mit Vorgabe der Stromtrajektorie durch off-line optimierte Pulsmuster (Fast Current Control for High Power Drives with Presetting of the Current Trajectory via Off-line Optimized Pulse Patterns)*, Dissertation, Wuppertal: Wuppertal University, 1998

[16] S. Bolognani, M. Zigliotto, Full-digital predictive hysteresis current control for switching losses minimisation in PMSM drives, *Power Electronics, Machines and Drives Conference PEMD'02*, pp. 61–67, Bath, 2002

[17] F. Bonanno, A. Consoli, A. Raciti, A. Testa, An Innovative Direct Self-Control Scheme for Induction Motor Drives, *IEEE Transactions on Power Electronics*, Vol. 12, No. 5, 1997, pp. 800–806

[18] C. Bordons, E. F. Camacho, A Generalized Predictive Controller for a Wide Class of Industrial Processes, *IEEE Transactions on Control Systems Technology*, Vol. 6, No. 3, 1998, pp. 372–387

[19] P. Boucher, D. Dumur, P. Raguenaud, C. Rougebief, Multivariable Generalised Predictive Control for Cognac Distillation, *European Control Conference ECC'99*, Karlsruhe, 1999

[20] E. F. Camacho, C. Bordons, *Model Predictive Control*, London: Springer-Verlag, 1999

[21] M. S. Carmeli, F. Castelli-Dezza, G. Superti-Furga, Smart Modulation: A new Approach to Power Converter Control, *European Conference on Power Electronics and Applications EPE 2001*, Graz, 2001

[22] D. Casadei, G. Serra, A. Tani, Implementation of a Direct Torque Control Algorithm for Induction Motors Based on Discrete Space Vector Modulation, *IEEE Transactions on Power Electronics*, Vol. 15, No. 4, 2000, pp. 769–777

[23] M. Chandorkar, New Techniques for Inverter Flux Control, *IEEE Transactions on Industry Applications*, Vol. 37, No. 3, 2001, pp. 880–887

[24] Y. A. Chapuis, C. Pelissou, D. Roye, Direct Torque Control of Induction Machine under Square Wave Conditions, *IEEE IAS Annual Meeting*, Vol. 1, pp. 343–349, Orlando, 1995

[25] S. Chattopadhyay, V. Ramanarayanan, V. Jayashankar, A Predictive Switching Modulator for Current Mode Control of High Power Factor Boost Rectifier, *IEEE Transactions on Power Electronics*, Vol. 18, No. 1, Part I, 2003, pp. 114–123

[26] J.-W. Choi, S.-K. Sul, New Current Control Concept—Minimum Time Current Control in the Three-Phase PWM Converter, *IEEE Transactions on Power Electronics*, Vol. 12, No. 1, 1997, pp. 124–131

[27] D. W. Clarke, C. Mohtadi, P. S. Tuffs, Generalized Predictive Control—Part I. The Basic Algorithm, *Automatica*, Vol. 23, No. 2, 1987, pp. 137–148

[28] D. W. Clarke, C. Mohtadi, P. S. Tuffs, Generalized Predictive Control—Part II. Extensions and Interpretations, *Automatica*, Vol. 23, No. 2, 1987, pp. 149–160

[29] D. W. Clarke, R. Scattolini, Constrained receding-horizon predictive control, *IEE Proceedings*, Vol. 138, Part D, No. 4, 1991, pp. 347–354

[30] D. W. Clarke, Adaptive Predictive Control, *Annual Review in Automatic Programming*, Vol. 20, 1996, pp. 83–94

[31] L. Cooper, M. W. Cooper, *Introduction to Dynamic Programming*, Oxford: Pergamon Press, 1981

[32] C. R. Cutler, B. L. Ramaker, Dynamic Matrix Control—A computer control algorithm, *Joint Automatic Control Conference 1980*, Vol. 1, pp. 806–811, San Francisco, 1980

[33] M. Depenbrock, Direkte Selbstregelung (DSR) für hochdynamische Drehfeldantriebe mit Stromrichterspeisung (Direct Self Control for Inverter Supplied High Dynamic AC Drives), *etzArchiv*, Vol. 7, 1985, pp. 211–218

[34] R. Dittmar, B.-M. Pfeiffer, *Modellbasierte prädiktive Regelung (Model Based Predictive Control)*, Munich, Vienna: R. Oldenbourg Verlag, 2004

[35] S. V. Emeljanov, *Automatische Regelsysteme mit veränderlicher Struktur (Automatic Control Systems with Variable Structure)*, Munich, Vienna: R. Oldenbourg Verlag, 1969

[36] J. Faßnacht, P. Mutschler, Direct Mean Torque Control with improved flux control, *European Conference on Power Electronics and Applications EPE 2003*, Toulouse, 2003

[37] E. Flach *Direkte Regelung des Drehmomentmittelwerts einer Induktionsmaschine (Direct Control of the Mean Torque Value of an Induction Machine*, Dissertation, Darmstadt: Technical University Darmstadt, 1999

[38] O. Föllinger, *Regelungstechnik (Control Technology)*, Heidelberg: Hüthig, 1994

[39] F. R. Gantmacher, *Matrizentheorie (Matrix Theory)*, Berlin, Heidelberg: Springer-Verlag, 1986

[40] C. E. García, D. M. Prett, M. Morari, Model Predictive Control: Theory and Practice—a Survey, *Automatica*, Vol. 25, No. 3, 1989, pp. 335–348

[41] H. P. Geering, *Berechnung von Zustandsraummodellen minimaler Ordnung aus der Übertragungsmatrix $G(s)$ (Calculation of State Space Models with Minimal Order from the Transfer Matrix $G(s)$)*, `http://www.imrtweb.ethz.ch/users/~geering/SGA1983v2.pdf`, Zurich: IMRT Press, 1999

[42] T. Geyer, F. D. Torrisi, M. Morari, Efficient Mode Enumeration of Compositional Hybrid Systems, *Hybrid Systems: Computation and Control, Lecture Notes in Computer Science*, Vol. 2623, April 2003, pp. 216–232

[43] T. Geyer, F. D. Torrisi, M. Morari, Optimal Complexity Reduction of Piecewise Affine Models based on Hyperplane Arrangements, *American Control Conference ACC2004*, Boston, 2004

[44] G. C. Goodwin, K. S. Sin, *Adaptive Filtering, Prediction and Control*, Englewood Cliffs: Prentice-Hall, 1984

[45] P. Grieder, M. Morari, Complexity Reduction of Receding Horizon Control, *IEEE Conference on Decision and Control CDC2003*, pp. 3179–3184, Maui, 2003

[46] P. Grieder, M. Kvasnica, M. Baotić, M. Morari, Low Complexity Control of Piecewise Affine Systems with Stability Guarantee, *American Control Conference ACC2004*, Boston, 2004

[47] G. Hadley, *Nichtlineare und dynamische Programmierung (Non-linear and Dynamic Programming)*, Würzburg: Physica-Verlag, 1969

[48] L. Harnefors, H.-P. Nee, Model-Based Current Control of AC Machines Using the Internal Model Control Method, *IEEE Transactions on Industry Applications*, Vol. 34, No. 1, 1998, pp. 133–141

[49] J. Hecht, *Design and Simulation of a New Predictive Current Control Strategy for DC Drives*, Diploma Thesis, Toronto, Karlsruhe: University of Toronto/Technical University Karlsruhe, 1991

[50] D. Hintze, *Asynchroner Vierquadranten-Drehstromantrieb mit Stromzwischenkreisumrichter und oberschwingungsarmen Maschinengrößen (Asynchronous Four-Quadrant AC Drive with Current Source Inverter and Machine Values with Low Harmonics)*, Dissertation, Munich: Technical University Munich, 1993

[51] U. Hoffmann, *Entwurf und Erprobung einer adaptiven Zweipunktregelung (Design and Proving of an Adaptive Two-Step Control)*, Dissertation, Aachen: Technial University Aachen, 1984

[52] U. Hoffmann, Eine adaptive Zweipunktregelung für Prozesse mit schaltenden Stellgliedern (An Adaptive Two-Step Control for Processes with Switching Actuators), *at – Automatisierungstechnik*, 35. Jahrgang, No. 5, 1987, pp. 184–191

[53] D. G. Holmes, D. A. Martin, Implementation of a Direct Digital Predictive Current Controller for Single and Three Phase Voltage Source Inverters, *IEEE IAS Annual Meeting*, Vol. 2, pp. 906–913, San Diego, 1996

[54] J. Holtz, U. Schwellenberg, A new Fast-Response Current Control Scheme for Line Controlled Converters, *International Semiconductor Power Converter Conference*, pp. 175–183, Orlando, 1982

[55] J. Holtz, S. Stadtfeld, A Predictive Controller for the Stator Current Vector of AC-Machines fed from a Switched Voltage Source, *International Power Electronics Conference IPEC*, Vol. 2, pp. 1665–1675, Tokio, 1983

[56] J. Holtz, The Dynamic Representation of AC Drive Systems by Complex Signal Flow Graphs, *International Symposium on Industrial Electronics ISIE'94*, pp. 1–6, Santiago de Chile, 1994

[57] J. Holtz, The Induction Motor—A Dynamic System, *IEEE IECON'94*, Vol. 1, pp. P1–P6, Bologna, 1994

[58] R. Isermann, *Digitale Regelsysteme (Digital Control Systems)*, Volume I, Berlin, Heidelberg: Springer-Verlag, 1987

[59] R. Isermann, *Digitale Regelsysteme (Digital Control Systems)*, Volume II, Berlin, Heidelberg: Springer-Verlag, 1987

[60] B.-J. Kang, C.-M. Liaw, A Robust Hysteresis Current-Controlled PWM Inverter for Linear PMSM Driven Magnetic Suspended Positioning System, *IEEE Transactions on Industrial Electronics*, Vol. 48, No. 5, 2001, pp. 956–967

[61] J.-K. Kang, S.-K. Sul, New Direct Torque Control of Induction Motor for Minimum Torque Ripple and Constant Switching Frequency, *IEEE Transactions on Industry Applications*, Vol. 35, No. 5, 1999, pp. 1076–1082

[62] P. P. Kanjilal, *Adaptive prediction and predictive control*, London: Peter Peregrinus, 1995

[63] A. Kaufmann, A. Henry-Labordère, *Integer and Mixed Programming: Theory and Applications*, New York, San Francisco, London: Academic Press, 1977

[64] M. P. Kaźmierkowski, M. A. Dzieniakowski, W. Sulkowski, Novel Space Vector Based Current Controllers for PWM-Inverters, *IEEE Transactions on Power Electronics*, Vol. 6, No. 1, 1991, pp. 158–166

[65] R. Kennel, *Prädiktives Führungsverfahren für Stromrichter (Predictive Control Strategy for Inverters)*, Dissertation, Kaiserslautern: Kaiserslautern University, 1984

[66] R. Kennel, A. Linder, M. Linke, Generalized Predictive Control (GPC)—Ready for Use in Drive Applications?, *32nd IEEE Power Electronics Specialists Conference pesc2001*, Vol. 4, pp. 1839–1844, Vancouver, 2001

[67] R. Kennel, A. Linder, Predictive Control for Electrical Drives—A Survey, *Korea-Germany Advanced Power Electronics Symposium 2001*, pp. 39–43, Seoul, 2001

[68] R. M. C. de Keyser, G. A. van de Velde, F. A. G. Dumortier, A Comparative Study of Self-adaptive Long-range Predictive Control Methods, *Automatica*, Vol. 24, No. 2, 1988, pp. 149–163

[69] H. Kohlmeier, O. Niermeyer, D. Schröder, High dynamic four-quadrant AC-motor drive with improved power-factor and on-line optimized pulse pattern with PROMC, *European Conference on Power Electronics and Applications EPE'85*, Vol. 3, pp. 3.173–3.178, Brüssel, 1985

[70] K. P. Kovács, I. Rácz, *Transiente Vorgänge in Wechselstrommaschinen (Transient Behavior of AC Machines)*, Volume I, Budapest: Akadémiai Kiadó, 1959

[71] K. P. Kovács, I. Rácz, *Transiente Vorgänge in Wechselstrommaschinen (Transient Behavior of AC Machines)*, Volume II, Budapest: Akadémiai Kiadó, 1959

[72] P. Krauss, K. Daß, T. Bünte, Prädiktiver Regler mit Kalman-Filter zur Zustandsschätzung (Predictive Controller with Kalman Filter for State Estimation), *at – Automatisierungstechnik*, Vol. 42, No. 12, 1994, pp. 533–539

[73] M. Kvasnica, P. Grieder, M. Baotić, M. Morari, Multi-Parametric Toolbox (MPT), *Hybrid Systems: Computation and Control, Lecture Notes in Computer Science*, Vol. 2993, March 2004, pp. 448–462

[74] K.-K. La, M.-H. Shin, D.-S. Hyun, Direct Torque Control of Induction Motor with Reduction of Torque Ripple, *IEEE IECON2000*, pp. 1087–1092, Nagoya, 2000

[75] P. Lancaster, M. Tismenetsky, *The Theory of Matrices*, Orlando: Academic Press, 1985

[76] C. Lascu, I. Boldea, F. Blaabjerg, A Modified Direct Torque Control (DTC) for Induction Motor Sensorless Drive, *IEEE IAS Annual Meeting*, Vol. 1, pp. 415–422, St. Louis, 1998

[77] C. Lascu, A. M. Trzynadlowski, Combining the Principles of Sliding Mode, Direct Torque Control and Space-Vector Modulation in a High-Performance Sensorless AC Drive, *IEEE Transactions on Industry Applications*, Vol. 40, No. 1, 2004, pp. 170–177

[78] J. B. Lasserre, An explicit equivalent positive semidefinite program for nonlinear 0-1 programs, *SIAM Journal on Optimization*, Vol. 12, No. 3, 2002, pp. 756–769

[79] W. Leonhard, *Control of Electrical Drives*, Berlin, Heidelberg: Springer-Verlag, 1985

[80] I. J. Leontaritis, S. A. Billings, Input-output parametric models for nonlinear systems Part I: deterministic non-linear systems, *International Journal of Control*, Vol. 41, No. 2, 1985, pp. 303–328

[81] I. J. Leontaritis, S. A. Billings, Input-output parametric models for nonlinear systems Part II: stochastic non-linear systems, *International Journal of Control*, Vol. 41, No. 2, 1985, pp. 329–344

[82] K. K. S. Leung, J. Y. C. Chiu, H. S. H. Chung, Boundary Control of Inverters Using Second-Order Switching Surface, *36th IEEE Power Electronics Specialists Conference pesc2005*, pp. 936–942, Recife, 2005

[83] P. C. Loh, D. G. Holmes, A Variable Band Universal Flux/Charge Modulator for VSI and CSI Modulation, *IEEE IAS Annual Meeting*, Chicago, 2001

[84] J. Lunze, *Regelungstechnik 1 (Control Technology 1)*, Berlin, Heidelberg: Springer-Verlag, 1996

[85] J. Lunze, *Regelungstechnik 2 (Control Technology 2)*, Berlin, Heidelberg: Springer-Verlag, 1997

[86] J. Maes, J. Melkebeek, Discrete Time Direct Torque Control of Induction Motors using Back-EMF measurement, *IEEE IAS Annual Meeting*, Vol. 1, pp. 407–414, St. Louis, 1998

[87] H. R. Mayer, G. Pfaff, Direct Control of Induction Motor Currents—Design and Experimental Results, *European Conference on Power Electronics and Applications EPE 1985*, Vol. 2, pp. 3.7–3.12, Brüssel, 1985

[88] M. Morari, J. H. Lee, Model predictive control: past, present and future, *Computers and Chemical Engineering*, Vol. 23, No. 4–5, 1999, pp. 667–682

[89] C. El Moucary, E. Mendes, T. Raithel, R. Ortega, A New Strategy for Direct Torque and Flux Control of Induction Motor Drives, *International Conference on Electrical Machines ICEM2000*, Vol. 2, pp. 923–927, Espoo, 2000

[90] P. Mutschler, Verfahren zur direkten Regelung der Geschwindigkeit eines elektrischen Antriebs (Principle for Direct Control of the Speed of an Electric Drive), *German Patent* DE 196 35 981 C 2, 1998

[91] P. Mutschler, M. Marcks, A Direct Control Method for Matrix Converters, *IEEE Transactions on Industrial Electronics*, Vol. 49, No. 2, 2002, pp. 362–369

[92] I. Nagy, Novel Adaptive Tolerance Band Based PWM for Field-Oriented Control of Induction Machines, *IEEE Transactions on Industrial Electronics*, Vol. 41, No. 4, 1994, pp. 406–417

[93] K. Nakano, T. Yamamoto, T. Hinamoto, A Design of Robust Self-Tuning GPC-Based PID Controllers, *IEEE IECON2003*, pp. 285–290, Roanoke, 2003

[94] M. E. Nillesen, J. L. Duarte, M. Pasquariello, A. Del Pizzo, Direct Torque Control with the Application of a Predictive Pulse Width Control, *IEEE IAS Annual Meeting*, Vol. 3, pp. 1375–1379, Rome, 2000

[95] T. Noguchi, M. Yamamoto, S. Kondo, I. Takahashi, Enlarging Switching Frequency in Direct Torque-Controlled Inverter by Means of Dithering, *IEEE Transactions on Industry Applications*, Vol. 35, No. 6, 1999, pp. 1358–1366

[96] T. Noguchi, H. Kodachi, I. Saito, High-Speed Current Vector Control of PWM Inverter Minimizing Current Error at Every Sampling Point, *IEEE IECON2002*, Sevilla, 2002

[97] I.-H. Oh, Y.-S. Jung, M.-J. Youn, A Source Voltage-Clamped Resonant Link Inverter for a PMSM Using a Predictive Current Control Technique, *IEEE Transactions on Power Electronics*, Vol. 14, No. 6, 1999, pp. 1122–1132

[98] G. Ose, G. Schiemann, H. Baumann, F. Stopp, *Ausgewählte Kapitel der Mathematik (Selected Units of Mathematics)*, Leipzig: VEB Fachbuchverlag, 1975

[99] G. Pfaff, A. Wick, Direkte Stromregelung bei Drehstromantrieben mit Pulswechselrichter (Direct Current Control of AC Drives with PWM Inverter), *rtp – Regelungstechnische Praxis*, Vol. 24, No. 11, 1983, pp. 472–477

[100] A. Purcell, P. P. Acarnley, Multilevel hysteresis comparator forms for direct torque control schemes, *Electronic Letters*, Vol. 34, No. 6, 1998, pp. 601–603

[101] A. Purcell, P. P. Acarnley, Enhanced Inverter Switching for Fast Response Direct Torque Control, *IEEE Transactions on Power Electronics*, Vol. 16, No. 3, 2001, pp. 382–389

[102] J. B. Rawlings, Tutorial Overview of Model Predictive Control, *IEEE Control Systems Magazine*, June 2000, pp. 38–52

[103] J. Rodríguez, J. Pontt, C. Silva, P. Cortés, U. Ammann, S. Rees, Predictive Direct Torque Control of an Induction Machine, *11th International Power Electronics and Motion Control Conference EPE-PEMC2004*, Riga, 2004

[104] S. Salama, T. Yehia, Verfahren zur verbesserten Führung eines Drehstrom-Pulswechselrichters (Principle for better Control of an Three-Phase PWM Inverter), *German Patent* DE 40 21 766 C 2, 1993

[105] P. Schmitz, *Entwurf und Erprobung einer adaptiven prädiktiven Dreipunktregelung (Design and Proving of an Adaptive Predictive Three-Step Control)*, Fortschritt-Berichte VDI Series 8, No. 210, Düsseldorf: VDI-Verlag, 1990

[106] M. Schoch, *Das Erweiterungsprinzip (The Expansion Strategy)*, Berlin: VEB Deutscher Verlag der Wissenschaften, 1976

[107] H. Schwarz, *Mehrfachregelungen (Multi-Dimensional Control)*, Volume 1, Berlin, Heidelberg: Springer-Verlag, 1967

[108] H. Schwarz, *Mehrfachregelungen (Multi-Dimensional Control)*, Volume 2, Berlin, Heidelberg: Springer-Verlag, 1971

[109] D. E. Seborg, The Prospects for Advanced Process Control, *10th IFAC World Congress*, Munich, 1987

[110] A. Sikorski, M. Korzeniewski, Improvement of Torque and Flux Control in DTC Method, *11th International Power Electronics and Motion Control Conference EPE-PEMC2004*, Riga, 2004

[111] A. Steimel, J. Wiesemann, Further Development of Direct Self Control for Application in Electric Traction, *International Symposium on Industrial Electronics ISIE'96*, Vol. 1, pp. 180–185, Warschau, 1996

[112] H. Stöcker, *DeskTop Stöcker: Taschenbuch mathematischer Formeln und moderner Verfahren (Pocket Book of Mathematical Formulas and Modern Methods)*, Thun: Harri Deutsch, 1999

[113] I. Takahashi, T. Noguchi, A New Quick Response and High Efficiency Control Strategy of an Induction Motor, *IEEE IAS Annual Meeting*, pp. 496–502, Toronto, 1985

[114] D. Telford, M. W. Dunnigan, B. W. Williams, A Novel Torque-Ripple Reduction Strategy for Direct Torque Control, *IEEE Transactions on Industrial Electronics*, Vol. 48, No. 4, 2001, pp. 867–870

[115] J. Terno, *Numerische Verfahren der diskreten Optimierung (Numerical Methods of Discrete Optimization)*, Leipzig: B. G. Teubner, 1981

[116] H. du Toit Mouton, J. H. R. Enslin, An Optimal On-Line-Tuning Current Regulator for High-Power IGBT Converters, *IEEE Transactions on Industry Applications*, Vol. 35, No. 5, 1999, pp. 1132–1140

[117] P. Tøndel, T. A. Johansen, A. Bemporad, Evaluation of piecewise affine control via binary search tree, *Automatica*, Vol. 39, No. 5, 2003, pp. 945–950

[118] A. M. Trzynadlowski, M. M. Bech, F. Blaabjerg, J. K. Pedersen, An Integral Space-Vector PWM Technique for DSP-Controlled Voltage-Source Inverters, *IEEE Transactions on Industry Applications*, Vol. 35, No. 5, 1999, pp. 1091–1097

[119] T. T. C. Tsang, D. W. Clarke, Generalised predictive control with input constraints, *IEE Proceedings*, Vol. 135, Part D, No. 6, 1988, pp. 451–460

[120] P. Vas, A. F. Stronach, M. Rashed, M. Zordan, B. C. Chew, DSP implementation of torque pulsation reduction techniques in sensorless DTC induction and P. M. synchronous motor drives, *Intelligent Motion Proceedings*, June 1999, pp. 107–111

[121] H. Warmer, D. Schröder, An improved method of predictive control for line commutated DC-drives, *International Conference on Electrical Machines ICEM'86*, pp. 566–571, Munich, 1986

[122] D. Wuest, F. Jenni, Space vector based current control schemes for voltage source inverters, *24th IEEE Power Electronics Specialists Conference pesc'93*, pp. 986–992, Seattle, 1993

[123] E. Zafiriou, M. Morari, Design of robust digital controllers and sampling-time selection for SISO systems, *International Journal of Control*, Vol. 44, No. 3, 1986. pp. 711–735

[124] R. Zurmühl, S. Falk, *Matrizen und ihre Anwendungen (Matrices and their Applications)*, Part 1: Basics, Berlin, Heidelberg: Springer-Verlag, 1984

[125] unknown, *Polynomial and polynomial matrix glossary*, PolyX Ltd., `http://www.polyx.cz/glossary.htm`, Prague, 2002

Appendix A

Glossary polynomial matrices

From [44] and [125]:

Polynomial matrix: A $k \times m$ polynomial matrix is a matrix of the form

$$\mathbf{P}(s) = \mathbf{P}_0 + \mathbf{P}_1 s + \mathbf{P}_2 s^2 + \cdots + \mathbf{P}_n s^n$$

in which s is an undefined, normally complex variable and

$$\mathbf{P}_0, \mathbf{P}_1, \mathbf{P}_2, \cdots \mathbf{P}_n$$

represents the constant $k \times m$ coefficient matrices. If nothing else is declared, *real* polynomial matrices, i. e. polynomial matrices whose coefficient matrices are real, are treated.

If $\mathbf{P}_n$ is not a zero matrix, then $\mathbf{P}(s)$ is of the degree n.

If $\mathbf{P}_n$ is an identity matrix, $\mathbf{P}(s)$ is said to be *monic*.

Tall and wide: A polynomial matrix is *tall*, if it has at least as many rows as columns. It is *wide*, if it has at least as many columns as rows.

Rank: A polynomial matrix $\mathbf{P}(s)$ has *full column rank* (or *full normal column rank*) if it has full column rank everywhere in the complex plane, except at a finite number of points. Similar definitions can be given for *full row rank* and *full rank*.

The *normal rank* of a polynomial matrix $\mathbf{P}(s)$ is equal to

$$\max_{s \in \mathbb{C}} \text{ rg } \mathbf{P}(s)$$

Similar definitions apply to the notions of *normal column rank* and *normal row rank*.

A square polynomial matrix is *nonsingular* if it has full normal rank for almost all complex values of s.

Row and column degrees: Let the elements of the $k \times m$ polynomial matrix $\mathbf{P}(s)$ be

$$P_{i,j}(s), \qquad i = 1, 2, \cdots, k \qquad j = 1, 2, \cdots, m$$

Then the numbers

$$\rho_i = \max_j \deg P_{i,j}(s), \qquad i = 1, 2, \cdots, k$$
$$\gamma_j = \max_i \deg P_{i,j}(s), \qquad j = 1, 2, \cdots, m$$

are the *row* and the *column degrees* of $\mathbf{P}(s)$, respectively.

Leading coefficient matrix: Supposing that the polynomial matrix $\mathbf{P}(s)$ has the row and column degrees

$$\rho_i, \quad i = 1, 2, \cdots, k$$
$$\gamma_j, \quad j = 1, 2, \cdots, m$$

the *leading column coefficient matrix* of $\mathbf{P}(s)$ is a constant matrix whose element $e_{i,j}$ is the coefficient of the term with power γ_j of the polynomial element $P_{i,j}(s)$ of the matrix $\mathbf{P}(s)$.

The *leading row coefficient matrix* of $\mathbf{P}(s)$ is a constant matrix whose element $e_{i,j}$ is the coefficient of the term with power ρ_j of the polynomial element $P_{i,j}(s)$ of the matrix $\mathbf{P}(s)$.

Column- and row-reduced: A polynomial matrix is *column reduced* if its leading column coefficient matrix has full column rank. It is *row reduced* if its leading row coefficient matrix has full row rank.

A nonsingular polynomial matrix $\mathbf{P}(s)$ is *row reduced* if, and only if, its leading row coefficient matrix is regular (non-singular). It is *column reduced* if, and only if, its leading column coefficient matrix is regular.

Any nonsingular polynomial matrix $\mathbf{P}(s)$ can be transformed into a row-reduced form by elementary row operations. Additionally, the resulting leading row coefficient matrix can be assumed to be a lower triangular matrix with unity entries on the main diagonal. In the same way $\mathbf{P}(s)$ can be converted into a column-reduced form by elementary column operations. Then, the resulting leading column coefficient matrix can assumed to be an upper triangular matrix with unity entries on the main diagonal.

Roots: The *roots* or *zeros* of a polynomial matrix $\mathbf{P}(s)$ are those points in the complex plane where $\mathbf{P}(s)$ loses rank.

If $\mathbf{P}(s)$ is square, its roots are the roots of its determinant $\det \mathbf{P}(s)$, including multiplicity.

Prime: A polynomial matrix $\mathbf{P}(s)$ is *left prime* if it has full row rank everywhere in the complex plane. The matrix is *right prime* if it has full column rank everywhere in the complex plane.

Coprime (relative prime): N polynomial matrices $\mathbf{P}_1(s), \mathbf{P}_2(s), \cdots \mathbf{P}_N(s)$ with the same number of rows are *left coprime* or *relatively left prime*, if

$$\begin{bmatrix} \mathbf{P}_1(s) & \mathbf{P}_2(s) & \cdots & \mathbf{P}_N(s) \end{bmatrix}$$

is left prime. If all N polynomial matrices have the same number of columns, then they are *right coprime* or *relatively right prime* if

$$\begin{bmatrix} \mathbf{P}_1(s) \\ \mathbf{P}_2(s) \\ \vdots \\ \mathbf{P}_N(s) \end{bmatrix}$$

is right prime.

Two matrices $\mathbf{D_L}(s)$ and $\mathbf{N_L}(s)$ with the same number of rows are *relatively left prime* or *left coprime* if, and only if, their greatest common left divisor is unimodular.

Two matrices $\mathbf{D_R}(s)$ and $\mathbf{N_R}(s)$ with the same number of columns are *relatively right prime* or *right coprime* if, and only if, their greatest common right divisor is unimodular.

Unimodular: A *unimodular matrix* $\mathbf{U}(s)$ is defined as an arbitrary square matrix which can be derived from the identity matrix by a finite number of elementary row and column operations. Therefore, the determinant $\det \mathbf{U}(s)$ is a nonzero real or complex scalar, and conversely any polynomial matrix whose determinant is a nonzero real or complex scalar is a unimodular matrix.

The inverse of a polynomial matrix, an *inverse matrix being in the ring area*, i. e. again a polynomial matrix, exists if, and only if, the original matrix is unimodular [124].

Elementary row and column operations: Three elementary row operations exist:

- Multiplication of a row with a real or complex-valued constant unequal to zero:

$$\begin{bmatrix} 1 & s \\ 2 & s^2 \end{bmatrix} \xrightarrow{\text{row } 1 \cdot 3} \begin{bmatrix} 3 & 3s \\ 2 & s^2 \end{bmatrix}$$

- Interchanging of two rows:

$$\begin{bmatrix} 1 & s \\ 2 & s^2 \end{bmatrix} \xrightarrow{\text{interchange row 1 and row 2}} \begin{bmatrix} 2 & s^2 \\ 1 & s \end{bmatrix}$$

- Addition of a polynomial multiple of one row to another one:

$$\begin{bmatrix} 1 & s \\ 2 & s^2 \end{bmatrix} \xrightarrow{\text{row } 1 = \text{row } 1 + s \cdot \text{row } 2} \begin{bmatrix} 1+2s & s+s^3 \\ 2 & s^2 \end{bmatrix}$$

Elementary column operations are defined analogously.

It shall be noted that the elementary row or column operations of a polynomial matrix $\mathbf{P}(s)$ can also be realized if identical operations are carried out on an identity matrix $\mathbf{I}$ and then $\mathbf{P}(s)$ is premultiplied with the result.

Diophantine equation: The most simple form of a linear, scalar polynomial equation—the so-called *Diophantine equation* named after the Alexandrinian mathematician Diophantos (A. D. 275)—is:

$$A(s)X(s) + B(s)Y(s) = C(s)$$

The polynomials $A(s)$, $B(s)$ and $C(s)$ are given, while the polynomials $X(s)$ and $Y(s)$ are unknown. The equation is solvable if, and only if, the greatest common divisor of $A(s)$ and $B(s)$ is also a divisor of $C(s)$. This implies that the equation is solvable for any right hand side polynomial including $C(s) = 1$, if $A(s)$ and $B(s)$ are coprime.

If a Diophantine equation is solvable, it always has an infinite number of solutions. If $(X_0(s), Y_0(s))$ is any (particular) solution, the general solution is

$$X(s) = X_0(s) + \overline{B}(s)T(s)$$
$$Y(s) = Y_0(s) - \overline{A}(s)T(s)$$

where $T(s)$ is an arbitrary polynomial (the parameter) and $\overline{A}(s)$ and $\overline{B}(s)$ are coprime polynomials to which the following applies:

$$\frac{\overline{B}(s)}{\overline{A}(s)} = \frac{B(s)}{A(s)}$$

If $A(s)$ and $B(s)$ themselves are coprime, then one can of course select

$$\overline{A}(s) = A(s), \qquad \overline{B}(s) = B(s)$$

Among all solutions of a Diophantine equation, a unique solution pair $(X(s), Y(s))$ exists, characterized by

$$\deg X(s) < \deg \overline{B}(s)$$

There is another—usually different—solution pair with

$$\deg Y(s) < \deg \overline{A}(s)$$

Both solution pairs are only identical if

$$\deg A(s) + \deg B(s) \geqq \deg C(s)$$

Bézout equation: A Diophantine equation with 1 on the right hand side is called *Bézout equation*. It could e. g. have the form

$$A(s)X(s) + B(s)Y(s) = 1$$

in which $A(s)$ and $B(s)$ are given polynomials and $X(s)$ and $Y(s)$ are unknown.

Two polynomial matrices $\mathbf{D_L}(s)$ and $\mathbf{N_L}(s)$ with the same number of rows are left coprime if, and only if, two polynomial matrices $\mathbf{A}(s)$ and $\mathbf{B}(s)$ exist, which fulfil the Bézout equation

$$\mathbf{D_L}(s)\mathbf{A}(s) + \mathbf{N_L}(s)\mathbf{B}(s) = \mathbf{I}$$

Two polynomial matrices $\mathbf{D_R}(s)$ and $\mathbf{N_R}(s)$ with the same number of columns are right coprime if, and only if, two polynomial matrices $\mathbf{A}(s)$ and $\mathbf{B}(s)$ exist, which fulfil the Bézout equation

$$\mathbf{A}(s)\mathbf{D_R}(s) + \mathbf{B}(s)\mathbf{N_R}(s) = \mathbf{I}$$

Divisors and multiples: Let the polynomials $A(s)$, $B(s)$ and $C(s)$ be given such that $A(s) = B(s) \cdot C(s)$. In this case, $B(s)$ is called divisor of $A(s)$ and $A(s)$ is called multiple of $B(s)$. It can be written as $B(s)|A(s)$. This is also stated as $B(s)$ divides $A(s)$.

If a polynomial $G(s)$ divides both $A(s)$ and $B(s)$, then $G(s)$ is called common divisor of $A(s)$ and $B(s)$. If additionally, $G(s)$ is a multiple of every common divisor of $A(s)$ and $B(s)$, then $G(s)$ is the greatest common divisor (GCD) of $A(s)$ and $B(s)$. If the only common divisors of $A(s)$ and $B(s)$ are constants, then $A(s)$ and $B(s)$ are coprime.

If a polynomial $M(s)$ is a multiple of both $A(s)$ and $B(s)$, then $M(s)$ is called common multiple of $A(s)$ and $B(s)$. If additionally, $M(s)$ is a divisor of all common multiples of $A(s)$ and $B(s)$, then $M(s)$ is the least common multiple (LCM) of $A(s)$ and $B(s)$.

Now let the polynomial matrices $\mathbf{A}(s)$, $\mathbf{B}(s)$ and $\mathbf{C}(s)$ of compatible size be given such that $\mathbf{A}(s) = \mathbf{B}(s) \cdot \mathbf{C}(s)$. In this case, $\mathbf{B}(s)$ is called the left divisor of $\mathbf{A}(s)$ and $\mathbf{A}(s)$ is called the right multiple of $\mathbf{B}(s)$.

If a polynomial matrix $\mathbf{G}(s)$ is a left divisor of $\mathbf{A}(s)$ and $\mathbf{B}(s)$, then $\mathbf{G}(s)$ is called common left divisor of $\mathbf{A}(s)$ and $\mathbf{B}(s)$. If additionally, $\mathbf{G}(s)$ is a right multiple of every common left divisor of $\mathbf{A}(s)$ and $\mathbf{B}(s)$, then $\mathbf{G}(s)$ is the greatest common left divisor of $\mathbf{A}(s)$ and $\mathbf{B}(s)$. If the only common left divisors of $\mathbf{A}(s)$ and $\mathbf{B}(s)$ are unimodular matrices, then the polynomial matrices $\mathbf{A}(s)$ and $\mathbf{B}(s)$ are left coprime.

If a polynomial matrix $\mathbf{M}(s)$ is a right multiple of $\mathbf{A}(s)$ and $\mathbf{B}(s)$, then $\mathbf{M}(s)$ is called common right multiple of $\mathbf{A}(s)$ and $\mathbf{B}(s)$. If additionally, $\mathbf{M}(s)$ is a left divisor of all common right multiples of $\mathbf{A}(s)$ and $\mathbf{B}(s)$, then $\mathbf{M}(s)$ is the smallest common right multiple of $\mathbf{A}(s)$ and $\mathbf{B}(s)$.

Right divisors, left multiples, common right divisors, greatest common right divisors, common left multiples and smallest common left multiples are defined in a similar way.

Other properties: For the sake of simplicity, the properties described here are either for left or for right matrix operations. The corresponding results for the other case can be derived by substituting right, row, column, premultiplication, etc. with left, column, row or postmultiplication.

Any $m \times l$ polynomial matrix $\mathbf{P}(s)$ of the rank r can be reduced by elementary column operations to a lower left triangular matrix in which the following applies:

- If $l > s$, then the last $(l - s)$ columns are all zero.
- In row j with $1 \leqq j \leqq s$, the diagonal element is monic and of a higher degree than any (nonzero) element at the left of it.
- If in row j with $1 \leqq j \leqq s$, the diagonal element is unity, then all elements at the left of it are zero.

The two $m \times m$ and $m \times r$ polynomial matrices $\mathbf{D_L}(s)$ and $\mathbf{N_L}(s)$ are given. If the matrix $\begin{bmatrix}\mathbf{D_L}(s) & \mathbf{N_L}(s)\end{bmatrix}$ is reduced to a lower left triangular form $\begin{bmatrix}\mathbf{R}(s) & 0\end{bmatrix}$ according to the above description, then $\mathbf{R}(s)$ is the greatest common left divisor of $\mathbf{D_L}(s)$ and $\mathbf{N_L}(s)$.

Appendix B

Nomenclature

The notation of the individual variables, parameters etc. is based on the following convention:

Scalar values	:	Italic font (a, b, c)
Complex values	:	Italic bold font $(\boldsymbol{a}, \boldsymbol{b}, \boldsymbol{c})$
Vectors	:	Bold lowercase letters $(\mathbf{a}, \mathbf{b}, \mathbf{c})$
Matrices	:	Bold capitals $(\mathbf{A}, \mathbf{B}, \mathbf{C})$
Sets	:	Italic, bold capitals $(\boldsymbol{A}, \boldsymbol{B}, \boldsymbol{C})$
Polynomials in z	:	Capitals $(A(z^{-1}), B(z^{-1}), C(z^{-1}))$
Polynomial matrices in z	:	Bold capitals $(\mathbf{A}(z^{-1}), \mathbf{B}(z^{-1}), \mathbf{C}(z^{-1}))$
Polytopes	:	Curved capitals $(\mathcal{A}, \mathcal{B}, \mathcal{C})$
Number ranges	:	Bold capitals with double line $(\mathbb{A}, \mathbb{B}, \mathbb{C})$
Reference values	:	Star superscript $(a^*, \boldsymbol{a}^*, \mathbf{a}^*)$

Overview of the formula symbols:

Machine values:

a	Acceleration
a	Switching state of the half bridge of phase a
b	Switching state of the half bridge of phase b
c	Switching state of the half bridge of phase c
d	Distortion factor
e	Control deviation, control error
f_s	Inverter switching frequency, non-normalized
I	Current (general)
I_R	Rated current
$I_{ph,R}$	Rated phase current
$\boldsymbol{i}_s$	Stator current
i_{sa}	Stator current in phase a

i_{sb}	Stator current in phase b
i_{sc}	Stator current in phase c
i_{sd}	Field-producing component of the stator current
i_{sq}	Torque-producing component of the stator current
$i_{s\alpha}$	Real part of $\boldsymbol{i}_s$ in stator coordinates
$i_{s\beta}$	Imaginary part of $\boldsymbol{i}_s$ in stator coordinates
k_r	Rotor coupling factor
k_s	Stator coupling factor
l_h	Mutual machine inductance
l_r	Rotor inductance
${l_r}'$	Transient rotor inductance
$l_{r\sigma}$	Leakage inductance of the rotor
l_s	Stator inductance
l_s'	Transient stator inductance
$l_{s\sigma}$	Leakage inductance of the stator
m	Modulation index
p	Number of pole pairs
r_r	Rotor resistance
r_s	Stator resistance
r_σ	Effective resistance of both windings
S	Inverter switching state
T_0	Sampling rate, non-normalized
U_R	Nominal machine voltage
$U_{ph,R}$	Nominal phase voltage
u_d	DC link voltage
$\boldsymbol{u}_s$	Stator voltage
u_{sd}	Field-producing component of the stator voltage
u_{sq}	Torque-producing component of the stator voltage
$u_{s\alpha}$	Real part of $\boldsymbol{u}_s$ in stator coordinates
$u_{s\beta}$	Imaginary part of $\boldsymbol{u}_s$ in stator coordinates
δ	Field angle
σ	Total leakage factor
τ_0	Sampling rate
τ_f	Filter time constant for the actual values
τ_m	Mechanical time constant
τ_r	Rotor time constant
${\tau_r}'$	Transient rotor time constant
τ_s	Stator time constant
${\tau_s}'$	Transient stator time constant

τ_{tot}	Dead time
$\tau_\sigma{}'$	Transient time constant
φ	Mechanical rotor position
$\boldsymbol{\psi}_r$,ψ_{rd}	Rotor flux
$\psi_{r\alpha}$	Real part of $\boldsymbol{\psi}_r$ in stator coordinates
$\psi_{r\beta}$	Imaginary part of $\boldsymbol{\psi}_r$ in stator coordinates
$\boldsymbol{\psi}_s$	Stator flux
ω	Mechanical rotor speed
ω_k	Angular speed of the coordinate system
ω_r	Angular speed of the rotor current
ω_s	Angular speed of the stator current
$\omega_{s,R}$	Nominal angular speed of the stator current
	PI controller constants
T_i	Integrator time constant of the current controllers
T_ω	Integrator time constant of the speed controller
V_i	Gain factor of the current controllers
V_ω	Gain factor of the speed controller
	MPC variables (SISO)
$A(z^{-1})$	Denominator polynomial of the transfer function $G(z^{-1})$
$\tilde{A}(z^{-1})$	Modified denominator polynomial; $\tilde{A}(z^{-1}) = \Delta A(z^{-1})$
$B(z^{-1})$	Numerator polynomial of the transfer function $G(z^{-1})$
$C(z^{-1})$	Noise polynomial
d	Discrete delay time
$E_j(z^{-1})$	Auxiliary polynomial
$F_j(z^{-1})$	Auxiliary polynomial
$\mathbf{F}(z^{-1})$	Transfer polynomial matrix for past controlled variables
$\mathbf{FG}'$	Aux. matrix for the calculation of the free response
$\mathbf{F\Gamma}$	Aux. matrix for the calc. of the free response with filt.
$f(t)$	Free response
$\mathbf{f}$	Vector of the free response $f(t)$
$f'(t)$	Free response with filtering
$\mathbf{f}'$	Vector of the free response with filtering
$\mathbf{G}$	Transfer matrix for the forced response
$G(z^{-1})$	Transfer function of the system
$\mathbf{G}'$	Transfer matrix for the forced response with filtering
$\mathbf{G}'(z^{-1})$	Transfer polynomial matrix for past actuating variables

$G_j(z^{-1})$	Transfer polynomial for GPC controller
$G_j{}'(z^{-1})$	Transfer function for future actuating variables
$\tilde{\mathbf{g}}^T$	First row of the matrix $(\mathbf{G}^T\mathbf{G} + \lambda\mathbf{I})^{-1} \cdot \mathbf{G}^T$
$\tilde{\mathbf{g}}'^T$	First row of the matrix $(\mathbf{G}'^T\mathbf{G}' + \lambda\mathbf{I})^{-1} \cdot \mathbf{G}'^T$
$\mathbf{I}$	Identity matrix
i	Counter variable
J	Cost function
j	Counter variable
k	Counter variable
N	Horizon (general)
N_1	Lower cost horizon
N_2	Upper cost horizon
N_p	Prediction horizon
N_u	Control horizon
na	Degree of the denominator polynomial $A(z^{-1})$
nb	Degree of the numerator polynomial $B(z^{-1})$
nc	Degree of the noise polynomial $C(z^{-1})$
nt	Degree of the design polynomial $T(z^{-1})$
r_j	Auxiliary values for the calculation of $E_j(z^{-1})$
$\tilde{R}(z^{-1})$	Auxiliary polynomial for the calculation of $E_j(z^{-1})$
$T(z^{-1})$	Design polynomial for filtering
$u(t)$	Actuating variable
$u^f(t)$	Actuating variable, filtered with $T(z^{-1})$
$\tilde{\mathbf{u}}$	Vector of actuating variable differences $\Delta u(t)$
$w(t)$	Reference variable
$\mathbf{w}$	Vector of reference variables $w(t)$
$y(t)$	Controlled variable
$y^f(t)$	Controlled variable, filtered with $T(z^{-1})$
$\hat{y}(t)$	Predicted value of the controlled variable
$\mathbf{y}$	Vector of the predicted controlled variables $\hat{y}(t)$
$\mathbf{yu}$	Vector of past controlled and actuating v. $y(t)$ and $u(t)$
$\mathbf{yu}^f$	Vector of filt. controlled and actuating v. $y^f(t)$ and $u^f(t)$
$\mathbf{\Gamma}(z^{-1})$	Transfer polynomial matrix for filtered actuating variables
$\mathit{\Gamma}_j(z^{-1})$	Transfer function for filtered actuating variables
Δ	Differential operator $1 - z^{-1}$
λ_j	Weighting factors for actuating variables
μ_j	Weighting factors for control errors
$\xi(t)$	Noise variable

MPC variables (MIMO)

$\mathbf{0}$	Zero matrix
$\mathbf{A}$	State matrix
$\mathbf{A}(z^{-1})$	Denominator polynomial matrix for the CARIMA model
$\tilde{\mathbf{A}}(z^{-1})$	Modified polynomial matrix; $\tilde{\mathbf{A}}(z^{-1}) = \Delta\mathbf{A}(z^{-1})$
$\mathbf{A_d}$	State matrix in the discrete-time model
$\mathbf{B}$	Input matrix
$\mathbf{B}(z^{-1})$	Numerator polynomial matrix for the CARIMA model
$\mathbf{B_d}$	Input matrix in the discrete-time model
$\mathbf{C}$	Output matrix
$\mathbf{C}(z^{-1})$	Noise polynomial matrix for the CARIMA model
$\mathbf{C_d}$	Output matrix in the discrete-time model
$\mathbf{D}$	Feedforward matrix
$\mathbf{D}(z^{-1})$	Noise polynomial matrix for the CARIMA model
$\mathbf{D_d}$	Feedforward matrix in the discrete-time model
$\mathbf{E_d}$	Noise input matrix in the discrete-time model
$\mathbf{E}_j(z^{-1})$	Auxiliary polynomial matrix
$\mathbf{F}$	Matrix of the free responses $\mathbf{f}(t)$
$\mathbf{F}'$	Matrix of the modified free responses $\mathbf{H}\tilde{\mathbf{V}} + \mathbf{F}$
$\mathbf{F}'$	Matrix of the free responses $\mathbf{f}'(t)$ (with filtering)
$\mathbf{F}''$	Matrix of the modified free resp. $\mathbf{H}\tilde{\mathbf{V}} + \mathbf{F}'$ (with filt.)
$\mathbf{F}_{N_{12}}$	Matrix $\mathbf{F}$ under consideration of the control horizon
$\mathbf{F}'_{N_{12}}$	Matrix $\mathbf{F}'$ under consideration of the control horizon
$\mathbf{F}''_{N_{12}}$	Matrix $\mathbf{F}''$ under consideration of the control horizon
$\mathbf{F}(z^{-1})$	Transfer polynomial matrix for past controlled variables
$\mathbf{F_d}$	Noise feedforward matrix in the discrete-time model
$\mathbf{F}_j(z^{-1})$	Auxiliary polynomial matrix
$\mathbf{FG_p}$	Auxiliary matrix for the calculation of the free response
$\mathbf{FG_p}_{N_{12}}$	Matrix $\mathbf{FG_p}$ under consideration of the control horizon
$\mathbf{FG_pH_p}$	Auxiliary matrix for the free response with noise
$\mathbf{FG_pH_p}_{N_{12}}$	Matrix $\mathbf{FG_pH_p}$ under cons. of the control horizon
$\mathbf{F\Gamma\Theta}$	Like $\mathbf{FG_pH_p}$, but with filtering
$\mathbf{F\Gamma\Theta}_{N_{12}}$	Matrix $\mathbf{F\Gamma\Theta}$ under consideration of the control horizon
$\mathbf{f}(t)$	Free response
$\mathbf{f}'(t)$	Free response with filtering
$\mathbf{f}_j$	Vector of the free response
$\mathbf{G}$	Transfer matrix for the forced response

$\mathbf{G}'$	Transfer matrix for the forced response with filtering
$\mathbf{G}_{N_{12u}}$	Matrix $\mathbf{G}$ under consideration of the control horizon
$\mathbf{G}'_{N_{12u}}$	Matrix $\mathbf{G}'$ under consideration of the control horizon
$\mathbf{G_d}(z^{-1})$	Discrete-time transfer matrix (matrix polynomial)
$\mathbf{G}_j(z^{-1})$	Transfer polynomial matrix for the forced response
$\mathbf{G}_j{}'(z^{-1})$	Like $\mathbf{G}_j(z^{-1})$, but with filtering
$\mathbf{G}_{jp}(z^{-1})$	Transfer polynomial matrix for past actuating variables
$\mathbf{G_p}(z^{-1})$	Transfer polynomial matrix for past actuating variables
$\mathbf{H}$	Transfer matrix for future disturbances
$\mathbf{H}'$	Like $\mathbf{H}$, but with filtering
$\mathbf{H}_{N_{12u}}$	Matrix $\mathbf{H}$ under consideration of the control horizon
$\mathbf{H}'_{N_{12u}}$	Matrix $\mathbf{H}'$ under consideration of the control horizon
$\mathbf{H_d}(z^{-1})$	Discrete-time dist. trans. matrix (matrix polynomial)
$\mathbf{H}_j(z^{-1})$	Transfer polynomial matrix for future disturbances
$\mathbf{H}_j{}'(z^{-1})$	Like $\mathbf{H}_j(z^{-1})$, but with filtering
$\mathbf{H}_{jp}(z^{-1})$	Transfer polynomial matrix for past disturbances
$\mathbf{H_p}(z^{-1})$	Transfer polynomial matrix for past disturbances
$\mathbf{HFG_pH_p}$	Aux. matrix for the calculation of the mod. free resp.
$\mathbf{H'F\Gamma\Theta}$	Like $\mathbf{HFG_pH_p}$, but with filtering
$\mathbf{I}$	Identity matrix
i	Counter variable
J	Cost function
j	Counter variable
$\mathbf{K}$	Controller matrix (LQR)
k	Counter variable
l	Number of disturbance variables
m	Number of system inputs
N	Horizon (general)
N_1	Lower cost horizon
N_2	Upper cost horizon
N_p	Prediction horizon
N_u	Control horizon
n	Number of system outputs
na	Degree of the denominator matrix polynomial $\mathbf{A}(z^{-1})$
nb	Degree of the numerator matrix polynomial $\mathbf{B}(z^{-1})$
nc	Degree of the noise matrix polynomial $\mathbf{C}(z^{-1})$
nd	Degree of the disturbance matrix polynomial $\mathbf{D}(z^{-1})$
nt	Degree of the design matrix polynomial $\mathbf{T}(z^{-1})$

$\mathbf{P}$	Weighting matrix (LQR)
$\mathbf{P}(z^{-1})$	Arbitrary polynomial matrix
$\mathbf{Q}$	Weighting matrix for the state variables
$\mathbf{R}$	Weighting matrix for the actuating variables
$\mathbf{R}_j$	Real auxiliary matrix for the calculation of $\mathbf{E}_j(z^{-1})$
$\tilde{\mathbf{R}}(z^{-1})$	Auxiliary polynomial matrix for the calculation of $\mathbf{E}_j(z^{-1})$
$\mathbf{S}$	Weighting matrix for the final value of the state variables
$\mathbf{T}(z^{-1})$	Design polynomial matrix for filtering
$\mathbf{U}(z)$	Z-transformed system input $\mathbf{u}(t)$
$\tilde{\mathbf{U}}$	Matrix of actuating variable differences $\Delta\mathbf{u}(t)$
$\tilde{\mathbf{U}}_{N_u}$	Matrix $\tilde{\mathbf{U}}$ under consideration of the control horizon
$\mathbf{u}(t)$	Actuating variables or systems inputs
$\mathbf{u}^f(t)$	Actuating variables, filtered with $\mathbf{T}(z^{-1})$
$\tilde{\mathbf{V}}$	Matrix of disturbance variable differences $\Delta\mathbf{v}(t)$
$\tilde{\mathbf{V}}\mathbf{YUV}$	Matrix of misc. future and past variable values
$\tilde{\mathbf{V}}\mathbf{YUV}^f$	Matrix of misc. future and past values with filter
$\mathbf{v}(t)$	Disturbance variables
$\mathbf{v}^f(t)$	Disturbance variables filtered with $\mathbf{T}(z^{-1})$
$\mathbf{W}$	Matrix of future reference variables $\mathbf{w}(t)$
$\mathbf{W}_{N_{12}}$	Matrix $\mathbf{W}$ under consideration of the control horizon
$\mathbf{w}(t)$	Reference variables
$\mathbf{x}(t)$	State variables
$\mathbf{Y}$	Matrix of the predicted controlled variables $\hat{\mathbf{y}}(t)$
$\mathbf{Y}_{N_{12}}$	Matrix $\mathbf{Y}$ under consideration of the control horizon
$\mathbf{Y}(z)$	Z-transformation of system output variables $\mathbf{y}(t)$
$\mathbf{YU}$	Matrix of past controlled and actuating v. $\mathbf{y}(t)$ and $\mathbf{u}(t)$
$\mathbf{YUV}$	Matrix of past controlled, actuating and dist. v.
$\mathbf{YUV}^f$	Matrix of past controlled, control and dist. v. with filt.
$\mathbf{y}(t)$	Controlled variables or system output variables
$\mathbf{y}^f(t)$	Controlled variables, filtered with $\mathbf{T}(z^{-1})$
$\hat{\mathbf{y}}(t)$	Predicted values of the controlled variables
$\mathbf{\Gamma}(z^{-1})$	Like $\mathbf{G}_\mathbf{p}(z^{-1})$, but with filtering
$\mathbf{\Gamma}_j(z^{-1})$	Like $\mathbf{G}_{jp}(z^{-1})$, but with filtering
Δ	Differential operator $1 - z^{-1}$
$\mathbf{\Theta}(z^{-1})$	Like $\mathbf{H}_\mathbf{p}(z^{-1})$, but with filtering
$\mathbf{\Theta}_j(z^{-1})$	Like $\mathbf{H}_{jp}(z^{-1})$, but with filtering
λ_j	Weighting factors for actuating variables
μ_j	Weighting factors for control errors

$\xi(t)$	Noise variable
	DMPC variables
$\mathbf{0}$	Zero matrix
$\mathbf{A}$	State matrix (discrete-time)
$\mathbf{A}_d$	State matrix (discrete-time), cons. delay
$\mathbf{A}_t$	State matrix (discrete-time), cons. tracking
$\mathbf{A}_{t,d}$	State matrix (discrete-time), cons. tr. and delay
$\mathbf{A}(z^{-1})$	Den. polynomial matrix for CARIMA model (GPC)
$\mathbf{a}_j$	Describing vector for hyperplane j
$\mathbf{B}$	Input matrix (discrete-time)
$\mathbf{B}_d$	Input matrix (discrete-time), cons. delay
$\mathbf{B}_t$	Input matrix (discrete-time), cons. tracking
$\mathbf{B}_{t,d}$	Input matrix (discrete-time), cons. tr. and delay
$\mathbf{B}(z^{-1})$	Num. polynomial matrix for CARIMA model (GPC)
$\mathbf{b}_j$	Describing vector for hyperplane j
$\mathbf{C}$	Output matrix (discrete-time)
$\mathbf{C}_d$	Output matrix (discrete-time), cons. delay
$\mathbf{C}_t$	Output matrix (discrete-time), cons. tracking
$\mathbf{C}_{t,d}$	Output matrix (discrete-time), cons. tr. and delay
$\mathbf{D}$	Feedforward matrix (discrete-time)
$\mathbf{D}_d$	Feedforward matrix (discrete-time), cons. delay
$\mathbf{D}_t$	Feedforward matrix (discrete-time), cons. tracking
$\mathbf{D}_{t,d}$	Feedforward matrix (discrete-time), cons. tr. and delay
d_j	Describing function for hyperplane j
$\mathbf{E}$	Matrix of constraints, modified cost function
$\mathbf{F}$	Weighting matrix, modified cost function
$\mathbf{F}$	Matrix of free responses $\mathbf{f}(t)$ (GPC)
$\boldsymbol{F}$	Set of indices of the control laws
$\mathbf{F}_i$	Matrix control law within polytope $\mathcal{P}_i$
F_k	Affine control law
$\mathbf{f}(t)$	Free responses (GPC)
$\mathbf{G}$	Matrix of constraints, modified cost function
$\mathbf{G}$	Transfer matrix for forced response (GPC)
$\mathbf{G}(z^{-1})$	Discrete-time transfer matrix (matrix polyn.) (GPC)
$\mathbf{G}_i$	Matrix control law within polytope $\mathcal{P}_i$
$\mathbf{H}$	Weighting matrix, modified cost function
$\mathbf{H}_i$	Describing matrix for polytope $\mathcal{P}_i$ in **HK** notation

$\mathbf{I}$	Identity matrix
$\boldsymbol{I}$	Set of indices i of polytopes $\mathcal{P}_i$
$\boldsymbol{I}_k$	Set of indices of all polytopes $\mathcal{P}_i$ being in $\mathcal{P}(\boldsymbol{J}_k)$
i	Counter variable
$iter$	Iteration step with minimum-time controller
J	Cost function
$\boldsymbol{J}$	Set of indices j of hyperplanes
J'	Modified cost function
J'_z	Modified cost function for explicit solution
J_1	Single-step cost function for minimum-time controller
$\boldsymbol{J}_k$	Result of the hyperplanes already evaluated in node N_k
$\boldsymbol{J}_x$	Set of indices of all separating hyperplanes rel. to $\mathcal{P}_x$
j	Counter variable
$\mathbf{K}$	Controller gain matrix
$\boldsymbol{K}$	Set of nodes N_k to be evaluated
$\mathbf{K}_i$	Describing matrix for the polytope $\mathcal{P}_i$ in **HK** notation
k	Counter variable
L	Number of hyperplanes
M	Marking for position of polytope rel. to the hyperpl.
$\boldsymbol{M}_x$	Set of markings M of the polytope $\mathcal{P}_x$
m	Number of system inputs
$\mathbb{N}$	Set of natural numbers
N_1	Lower cost horizon
N_2	Upper cost horizon
N_k	Node k of the binary search tree
N_p	Prediction horizon
N_u	Control horizon
n	Number of the system states
na	Degree of the den. matrix polynomial $\mathbf{A}(z^{-1})$ (GPC)
nb	Degree of the num. matrix polynomial $\mathbf{B}(z^{-1})$ (GPC)
nt	Degree of the design matrix polynomial $\mathbf{T}(z^{-1})$ (GPC)
P	Number of polytopes of the explicit solution
$\mathbf{P}$	Weight. matrix for the final value of the state variables
$\mathcal{P}_j$	Polytope of the explicit solution
p	Number of system outputs
Q	Number of polytopes of the reduced explicit solution
$\mathbf{Q}$	Weighting matrix for the state variables
$\mathbf{Q}_d$	Like $\mathbf{Q}$, but under consideration of delay

$\mathbf{Q}_t$	Like $\mathbf{Q}$, but under consideration of tracking
$\mathbf{Q}_{t,d}$	Like $\mathbf{Q}$, but under consideration of tr. and delay
$\mathcal{Q}_j$	Polytope of the reduced explicit solution
R	Number of polytopes in a region with min-time contr.
$\mathbf{R}$	Weighting matrix for the actuating variables
$\mathbf{R}_d$	Like $\mathbf{R}$, but under consideration of delay
$\mathbf{R}_t$	Like $\mathbf{R}$, but under consideration of tracking
$\mathbf{R}_{t,d}$	Like $\mathbf{R}$, but under consideration of tracking and delay
$\mathbf{R}_\triangle$	Weighting matrix for changes of the actuating variables
$\mathbb{R}$	Set of real numbers
$\boldsymbol{R}$	Arbitrary set
$\mathbf{S}$	Matrix for constraints for the explicit solution
$\boldsymbol{S}$	Arbitrary set
s	Dimension of $\mathbf{U}$
$\mathbf{T}(z^{-1})$	Design polynomial matrix with filter (GPC)
$\mathcal{T}_{set}$	Target area for minimum-time controller
$\mathbf{U}$	Vector of all actuating variables $\mathbf{u}$
$\tilde{\mathbf{U}}$	Matrix of actuating variable differences $\Delta\mathbf{u}(t)$ (GPC)
$\boldsymbol{U}$	Set of possible values for the actuating variables
$\boldsymbol{U}_k$	Subset for the expansion principle
$\mathbf{u}, \mathbf{u}(t)$	Input vector, actuating variables
$\mathbf{u}_d, \mathbf{u}_d(t)$	Input vector, actuating variables, cons. delay
$\mathbf{u}_t, \mathbf{u}_t(t)$	Input vector, actuating variables, cons. tracking
$\mathbf{u}_{t,d}, \mathbf{u}_{t,d}(t)$	Input vector, actuating variables, cons. tr. and delay
$\mathbf{W}$	Matrix for constraints, modified cost function
$\mathbf{W}$	Matrix of future reference variables $\mathbf{w}(t)$ (GPC)
$\mathbf{w}, \mathbf{w}(t)$	Reference variables
$\mathbf{w}_d, \mathbf{w}_d(t)$	Reference variables, cons. of delay
$\mathbf{w}_t, \mathbf{w}_t(t)$	Reference variables, cons. of tracking
$\mathbf{w}_{t,d}, \mathbf{w}_{t,d}(t)$	Reference variables, cons. of tracking and delay
$\boldsymbol{X}$	Set of possible values for the state variables
$\mathcal{X}_{LQR}$	Central region with LQR control for min-time controllers
$\mathcal{X}_f^N$	Feasible region for N sampling cycles
$\mathbf{x}, \mathbf{x}(t)$	State vector
$\mathbf{x}_d, \mathbf{x}_d(t)$	State vector, cons. of delay
$\mathbf{x}_t, \mathbf{x}_t(t)$	State vector, cons. of tracking
$\mathbf{x}_{t,d}, \mathbf{x}_{t,d}(t)$	State vector, cons. of tracking and delay
$\mathbf{Y}$	Weighting matrix, modified cost function

$\boldsymbol{Y}$	Set of possible values for the controlled variables
$\mathbf{y}, \mathbf{y}(t)$	Output vector, controlled variables
$\mathbf{y}_d, \mathbf{y}_d(t)$	Output vector, controlled variables, cons. of delay
$\mathbf{y}_t, \mathbf{y}_t(t)$	Output vector, controlled variables, cons. of tracking
$\mathbf{y}_{t,d}, \mathbf{y}_{t,d}(t)$	Source vector, controlled variables, cons. of tr. and delay
$\mathbf{z}$	Auxiliary vector for the explicit solution
λ	Weighting factor for the actuating variables

Control Engineering (general)

$\mathbf{A}$	State matrix
$\mathbf{A_d}$	State matrix in a discrete-time model
$\mathbf{B}$	Input matrix
$\mathbf{B_d}$	Input matrix in a discrete-time model
$\mathbf{C}$	Output matrix
$\mathbf{C_d}$	Output matrix in a discrete-time model
$\mathbf{D}$	Feedforward matrix
$\mathbf{D_d}$	Feedforward matrix in a discrete-time model
$\mathbf{E}$	Input matrix for disturbance variables
$\mathbf{E_d}$	Disturbance input matrix in a discrete-time model
$\mathbf{F}$	Feedforward matrix for disturbance variables
$\mathbf{F_d}$	Disturbance feedf. matrix in a discrete-time model
$F_R(s)$	Controller transfer function
$\mathbf{I}$	Identity matrix
k	Discrete-time point at time kt
$\mathbf{k}$	Arbitrary constant vector
t	Time (general)
T_0	Sampling time
$\mathbf{u}$	Input vector
$\mathbf{v}$	Input vector of the disturbance variables
$\mathbf{x}$	State vector
$\mathbf{y}$	Output vector
τ	Normalized time
$\mathbf{\Phi}$	Transition or fundamental matrix

Matrix algebra

$\mathbf{A}$	Arbitrary matrix
$\mathbf{B}$	Auxiliary matrix for intermediate results
$\mathbf{I}$	Identity matrix
$\mathbf{L}$	Left (lower) triangular matrix

$\mathbf{P}$	Permutation matrix
$\mathbf{R}$	Right (upper) triangular matrix
$\mathbf{x}, \mathbf{y}$	Arbitrary vectors

Appendix C

Normalization values

The nominal phase values are used for normalization.

Star connection	$U_{ph,R} = \frac{1}{\sqrt{3}} U_R$ $I_{ph,R} = I_R$
Y-connection	$U_{ph,R} = U_R$ $I_{ph,R} = \frac{1}{\sqrt{3}} I_R$

Thereby the following normalization values result:

Voltage	$\sqrt{2} \cdot U_{ph,R}$
Current	$\sqrt{2} \cdot I_{ph,R}$
Impedance	$\frac{U_{ph,R}}{I_{ph,R}}$
Inductance	$\frac{U_{ph,R}}{\omega_{s,R} \cdot I_{ph,R}}$

Flux linkage	$\dfrac{\sqrt{2} \cdot U_{ph,R}}{\omega_{s,R}}$
Power	$3 \cdot U_{ph,R} \cdot I_{ph,R}$
Torque	$\dfrac{3 \cdot p \cdot U_{ph,R} \cdot I_{ph,R}}{\omega_{s,R}}$
Rotating speed	$\dfrac{\omega_{s,R}}{p}$
Time	$\dfrac{1}{\omega_{s,R}}$

Appendix D

Physical machine constants

Technical data of the asynchronous machine used in the experiments in the chapters 3.4, 6.1.2, 6.3.3, 7.2, 8.4, 9.4.3 and 9.5.5:

Machine constant	value
U_R	380 V
I_R	4.9 A
$\omega_{s,R}$	50 Hz
l_s	2.67
τ_r	95.2
r_s	0.0447
τ_s	59.7
$l_{s\sigma} = l_{r\sigma}$	0.0863
$l_h = l_s + l_{s\sigma}$	2.58
$r_r = \frac{l_r}{\tau_r}$	0.0280
σ	0.0663
$l_s{}' = l_r{}'$	0.177
$k_r = k_s$	0.966
$\tau_s{}'$	3.96
r_σ	0.0708
$\tau_\sigma{}'$	2.50
τ_m	550

Appendix E

Polynomials and matrices for GPC

E.1 SISO system

$$
\begin{aligned}
A(z^{-1}) &= 1 + a_1 z^{-1} + a_2 z^{-2} + \cdots + a_{na} z^{-na} \\
B(z^{-1}) &= b_0 + b_1 z^{-1} + b_2 z^{-2} + \cdots + b_{nb} z^{-nb} \\
C(z^{-1}) &= 1 + c_1 z^{-1} + c_2 z^{-2} + \cdots + c_{nc} z^{-nc} \\
E_j(z^{-1}) &= e_{j,0} + e_{j,1} z^{-1} + e_{j,2} z^{-2} + \cdots + e_{j,j-1} z^{-(j-1)} \qquad j = 1 \dots N_p \\
&= e_0 + e_1 z^{-1} + e_2 z^{-2} + \cdots + e_{j-1} z^{-(j-1)} \qquad \text{(see chapter 6.1.1)} \\
F_j(z^{-1}) &= f_{j,0} + f_{j,1} z^{-1} + f_{j,2} z^{-2} + \cdots + f_{j,na} z^{-na} \qquad j = 1 \dots N_p \\
\mathbf{F}(z^{-1}) &= \begin{bmatrix} F_1(z^{-1}) \\ F_2(z^{-1}) \\ \vdots \\ F_{N_p}(z^{-1}) \end{bmatrix} = \begin{bmatrix} f_{1,0} + f_{1,1} z^{-1} + \cdots + f_{1,na} z^{-na} \\ f_{2,0} + f_{2,1} z^{-1} + \cdots + f_{2,na} z^{-na} \\ \vdots \\ f_{N_p,0} + f_{N_p,1} z^{-1} + \cdots + f_{N_p,na} z^{-na} \end{bmatrix} \\
\mathbf{FG}' &= \begin{bmatrix} f_{1,0} & f_{1,1} & \cdots & f_{1,na} & g'_{1,0} & g'_{1,1} & \cdots & g'_{1,nb-1} \\ f_{2,0} & f_{2,1} & \cdots & f_{2,na} & g'_{2,0} & g'_{2,1} & \cdots & g'_{2,nb-1} \\ \vdots & \vdots & \vdots & \vdots & \vdots & \vdots & \vdots & \vdots \\ f_{N_p,0} & f_{N_p,1} & \cdots & f_{N_p,na} & g'_{N_p,0} & g'_{N_p,1} & \cdots & g'_{N_p,nb-1} \end{bmatrix} \\
\mathbf{F\Gamma} &= \begin{bmatrix} f_{1,0} & f_{1,1} & \cdots & f_{1,na} & \gamma_{1,0} & \gamma_{1,1} & \cdots & \gamma_{1,nb-1} \\ f_{2,0} & f_{2,1} & \cdots & f_{2,na} & \gamma_{2,0} & \gamma_{2,1} & \cdots & \gamma_{2,nb-1} \\ \vdots & \vdots & \vdots & \vdots & \vdots & \vdots & \vdots & \vdots \\ f_{N_p,0} & f_{N_p,1} & \cdots & f_{N_p,na} & \gamma_{N_p,0} & \gamma_{N_p,1} & \cdots & \gamma_{N_p,nb-1} \end{bmatrix}
\end{aligned}
$$

$$\mathbf{f} = \begin{bmatrix} f(t+1) \\ f(t+2) \\ \vdots \\ f(t+N_p) \end{bmatrix}$$

$$\mathbf{f}' = \begin{bmatrix} f'(t+1) \\ f'(t+2) \\ \vdots \\ f'(t+N_p) \end{bmatrix}$$

$$\mathbf{G} = \begin{bmatrix} g_0 & 0 & \cdots & 0 \\ g_1 & g_0 & \cdots & 0 \\ \vdots & \vdots & \ddots & \vdots \\ g_{N_p-1} & g_{N_p-2} & \cdots & g_0 \end{bmatrix}$$

$$\mathbf{G}' = \begin{bmatrix} {g'}_0 & 0 & 0 & \cdots & 0 \\ {g'}_1 & {g'}_0 & 0 & \cdots & 0 \\ {g'}_2 & {g'}_1 & {g'}_0 & \cdots & 0 \\ \vdots & \vdots & \vdots & \ddots & \vdots \\ {g'}_{N_p-1} & {g'}_{N_p-2} & {g'}_{N_p-3} & \cdots & {g'}_0 \end{bmatrix}$$

$$\mathbf{G}'(z^{-1}) = \begin{bmatrix} {g'}_{1,0} + {g'}_{1,1} z^{-1} + \cdots + {g'}_{1,nb-1} z^{-(nb-1)} \\ {g'}_{2,0} + {g'}_{2,1} z^{-1} + \cdots + {g'}_{2,nb-1} z^{-(nb-1)} \\ \vdots \\ {g'}_{N_p,0} + {g'}_{N_p,1} z^{-1} + \cdots + {g'}_{N_p,nb-1} z^{-(nb-1)} \end{bmatrix}$$

$$T(z^{-1}) = 1 + t_1 z^{-1} + t_2 z^{-2} + \cdots + t_{nt} z^{-nt}$$

$$\tilde{\mathbf{u}} = \begin{bmatrix} \Delta u(t) \\ \Delta u(t+1) \\ \vdots \\ \Delta u(t+N_p-1) \end{bmatrix}$$

$$\mathbf{w} = \begin{bmatrix} w(t+1) \\ w(t+2) \\ \vdots \\ w(t+N_p) \end{bmatrix}$$

$$\mathbf{y} = \begin{bmatrix} \hat{y}(t+1) \\ \hat{y}(t+2) \\ \vdots \\ \hat{y}(t+N_p) \end{bmatrix}$$

$$\mathbf{yu} = \begin{bmatrix} y(t) \\ y(t-1) \\ \cdots \\ y(t-na) \\ \Delta u(t-1) \\ \Delta u(t-2) \\ \cdots \\ \Delta u(t-nb) \end{bmatrix}$$

$$\mathbf{yu}^f = \begin{bmatrix} y^f(t) \\ y^f(t-1) \\ \cdots \\ y^f(t-na) \\ \Delta u^f(t-1) \\ \Delta u^f(t-2) \\ \cdots \\ \Delta u^f(t-nb) \end{bmatrix}$$

$$\mathbf{\Gamma}(z^{-1}) = \begin{bmatrix} \Gamma_1(z^{-1}) \\ \Gamma_2(z^{-1}) \\ \vdots \\ \Gamma_{N_p}(z^{-1}) \end{bmatrix} = \begin{bmatrix} \gamma_{1,0} + \gamma_{1,1}z^{-1} + \cdots + \gamma_{1,nb-1}z^{-(nb-1)} \\ \gamma_{2,0} + \gamma_{2,1}z^{-1} + \cdots + \gamma_{2,nb-1}z^{-(nb-1)} \\ \vdots \\ \gamma_{N_p,0} + \gamma_{N_p,1}z^{-1} + \cdots + \gamma_{N_p,nb-1}z^{-(nb-1)} \end{bmatrix}$$

E.2 MIMO system

E.2.1 Definitions

$$\begin{aligned}
\mathbf{A}(z^{-1}) &= \mathbf{I} + \mathbf{A}_1 z^{-1} + \mathbf{A}_2 z^{-2} + \cdots + \mathbf{A}_{na} z^{-na} \\
\mathbf{B}(z^{-1}) &= \mathbf{B}_0 + \mathbf{B}_1 z^{-1} + \mathbf{B}_2 z^{-2} + \cdots + \mathbf{B}_{nb} z^{-nb} \\
\mathbf{C}(z^{-1}) &= \mathbf{I} + \mathbf{C}_1 z^{-1} + \mathbf{C}_2 z^{-2} + \cdots + \mathbf{C}_{nc} z^{-nc} \\
\mathbf{D}(z^{-1}) &= \mathbf{D}_0 + \mathbf{D}_1 z^{-1} + \mathbf{D}_2 z^{-2} + \cdots + \mathbf{D}_{nd} z^{-nd} \\
\mathbf{E}_j(z^{-1}) &= \mathbf{E}_{j,0} + \mathbf{E}_{j,1} z^{-1} + \mathbf{E}_{j,2} z^{-2} + \cdots + \mathbf{E}_{j,j-1} z^{-(j-1)} \qquad j = 1 \ldots N_p
\end{aligned}$$

$$= \mathbf{E}_0 + \mathbf{E}_1 z^{-1} + \mathbf{E}_2 z^{-2} + \cdots + \mathbf{E}_{j-1} z^{-(j-1)} \qquad \text{(s. chapter 8.1.3)}$$

$$\mathbf{F} = \begin{bmatrix} \mathbf{f}(t+1) \\ \mathbf{f}(t+2) \\ \vdots \\ \mathbf{f}(t+N_p) \end{bmatrix}$$

$\mathbf{F}'$ and $\mathbf{F}''$ are defined accordingly via $\mathbf{f}'(t)$ and $\mathbf{f}''(t)$

$$\mathbf{F}_{N_{12}} = \begin{bmatrix} \mathbf{f}(t+N_1) \\ \mathbf{f}(t+N_1+1) \\ \vdots \\ \mathbf{f}(t+N_2) \end{bmatrix}$$

${\mathbf{F}'}_{N_{12}}$ and ${\mathbf{F}''}_{N_{12}}$ are defined accordingly via $\mathbf{f}'(t)$ and $\mathbf{f}''(t)$

$$\mathbf{F}(z^{-1}) = \begin{bmatrix} \mathbf{F}_1(z^{-1}) \\ \mathbf{F}_2(z^{-1}) \\ \vdots \\ \mathbf{F}_{N_p}(z^{-1}) \end{bmatrix} = \begin{bmatrix} \mathbf{F}_{1,0} + \mathbf{F}_{1,1} z^{-1} + \cdots + \mathbf{F}_{1,na} z^{-na} \\ \mathbf{F}_{2,0} + \mathbf{F}_{2,1} z^{-1} + \cdots + \mathbf{F}_{2,na} z^{-na} \\ \vdots \\ \mathbf{F}_{N_p,0} + \mathbf{F}_{N_p,1} z^{-1} + \cdots + \mathbf{F}_{N_p,na} z^{-na} \end{bmatrix}$$

$$\mathbf{F}_j(z^{-1}) = \mathbf{F}_{j,0} + \mathbf{F}_{j,1} z^{-1} + \mathbf{F}_{j,2} z^{-2} + \cdots + \mathbf{F}_{j,na} z^{-na} \qquad j = 1 \ldots N_p$$

$$\mathbf{FG_p} = \begin{bmatrix} \mathbf{F}_{1,0} & \cdots & \mathbf{F}_{1,na} & \mathbf{G}_{1p,0} & \cdots & \mathbf{G}_{1p,nb-1} \\ \mathbf{F}_{2,0} & \cdots & \mathbf{F}_{2,na} & \mathbf{G}_{2p,0} & \cdots & \mathbf{G}_{2p,nb-1} \\ \vdots & \vdots & \vdots & \vdots & \vdots & \vdots \\ \mathbf{F}_{N_p,0} & \cdots & \mathbf{F}_{N_p,na} & \mathbf{G}_{N_p p,0} & \cdots & \mathbf{G}_{N_p p,nb-1} \end{bmatrix}$$

$$\mathbf{FG_p}_{N_{12}} = \begin{bmatrix} \mathbf{F}_{N_1,0} & \cdots & \mathbf{F}_{N_1,na} & \mathbf{G}_{N_1 p,0} & \cdots & \mathbf{G}_{N_1 p,nb-1} \\ \mathbf{F}_{N_1+1,0} & \cdots & \mathbf{F}_{N_1+1,na} & \mathbf{G}_{(N_1+1)p,0} & \cdots & \mathbf{G}_{(N_1+1)p,nb-1} \\ \vdots & \vdots & \vdots & \vdots & \vdots & \vdots \\ \mathbf{F}_{N_2,0} & \cdots & \mathbf{F}_{N_2,na} & \mathbf{G}_{N_2 p,0} & \cdots & \mathbf{G}_{N_2 p,nb-1} \end{bmatrix}$$

$$\mathbf{FG_pH_p} = \begin{bmatrix} \mathbf{F}_{1,0} & \mathbf{F}_{1,1} & \cdots & \mathbf{F}_{1,na} \\ \mathbf{F}_{2,0} & \mathbf{F}_{2,1} & \cdots & \mathbf{F}_{2,na} \\ \vdots & \vdots & \vdots & \vdots \\ \mathbf{F}_{N_p,0} & \mathbf{F}_{N_p,1} & \cdots & \mathbf{F}_{N_p,na} \\ \mathbf{G}_{1p,0} & \mathbf{G}_{1p,1} & \cdots & \mathbf{G}_{1p,nb-1} \\ \mathbf{G}_{2p,0} & \mathbf{G}_{2p,1} & \cdots & \mathbf{G}_{2p,nb-1} \\ \vdots & \vdots & \vdots & \vdots \\ \mathbf{G}_{N_pp,0} & \mathbf{G}_{N_pp,1} & \cdots & \mathbf{G}_{N_pp,nb-1} \\ \mathbf{H}_{1p,0} & \mathbf{H}_{1p,1} & \cdots & \mathbf{H}_{1p,nd-1} \\ \mathbf{H}_{2p,0} & \mathbf{H}_{2p,1} & \cdots & \mathbf{H}_{2p,nd-1} \\ \vdots & \vdots & \vdots & \vdots \\ \mathbf{H}_{N_pp,0} & \mathbf{H}_{N_pp,1} & \cdots & \mathbf{H}_{N_pp,nd-1} \end{bmatrix}$$

$$\mathbf{FG_pH_p}_{N_{12}} = \begin{bmatrix} \mathbf{F}_{N_1,0} & \mathbf{F}_{N_1,1} & \cdots & \mathbf{F}_{N_1,na} \\ \mathbf{F}_{N_1+1,0} & \mathbf{F}_{N_1+1,1} & \cdots & \mathbf{F}_{N_1+1,na} \\ \vdots & \vdots & \vdots & \vdots \\ \mathbf{F}_{N_2,0} & \mathbf{F}_{N_2,1} & \cdots & \mathbf{F}_{N_2,na} \\ \mathbf{G}_{N_1p,0} & \mathbf{G}_{N_1p,1} & \cdots & \mathbf{G}_{N_1p,nb-1} \\ \mathbf{G}_{(N_1+1)p,0} & \mathbf{G}_{(N_1+1)p,1} & \cdots & \mathbf{G}_{(N_1+1)p,nb-1} \\ \vdots & \vdots & \vdots & \vdots \\ \mathbf{G}_{N_2p,0} & \mathbf{G}_{N_2p,1} & \cdots & \mathbf{G}_{N_2p,nb-1} \\ \mathbf{H}_{N_1p,0} & \mathbf{H}_{N_1p,1} & \cdots & \mathbf{H}_{N_1p,nd-1} \\ \mathbf{H}_{(N_1+1)p,0} & \mathbf{H}_{(N_1+1)p,1} & \cdots & \mathbf{H}_{(N_1+1)p,nd-1} \\ \vdots & \vdots & \vdots & \vdots \\ \mathbf{H}_{N_2p,0} & \mathbf{H}_{N_2p,1} & \cdots & \mathbf{H}_{N_2p,nd-1} \end{bmatrix}$$

$$\mathbf{F\Gamma\Theta} = \begin{bmatrix} \mathbf{F}_{1,0} & \mathbf{F}_{1,1} & \cdots & \mathbf{F}_{1,na} \\ \mathbf{F}_{2,0} & \mathbf{F}_{2,1} & \cdots & \mathbf{F}_{2,na} \\ \vdots & \vdots & \vdots & \vdots \\ \mathbf{F}_{N_p,0} & \mathbf{F}_{N_p,1} & \cdots & \mathbf{F}_{N_p,na} \\ \mathbf{\Gamma}_{1,0} & \mathbf{\Gamma}_{1,1} & \cdots & \mathbf{\Gamma}_{1,nb-1} \\ \mathbf{\Gamma}_{2,0} & \mathbf{\Gamma}_{2,1} & \cdots & \mathbf{\Gamma}_{2,nb-1} \\ \vdots & \vdots & \vdots & \vdots \\ \mathbf{\Gamma}_{N_p,0} & \mathbf{\Gamma}_{N_p,1} & \cdots & \mathbf{\Gamma}_{N_p,nb-1} \\ \mathbf{\Theta}_{1,0} & \mathbf{\Theta}_{1,1} & \cdots & \mathbf{\Theta}_{1,nd-1} \\ \mathbf{\Theta}_{2,0} & \mathbf{\Theta}_{2,1} & \cdots & \mathbf{\Theta}_{2,nd-1} \\ \vdots & \vdots & \vdots & \vdots \\ \mathbf{\Theta}_{N_p,0} & \mathbf{\Theta}_{N_p,1} & \cdots & \mathbf{\Theta}_{N_p,nd-1} \end{bmatrix}$$

$$\mathbf{F\Gamma\Theta}_{N_{12}} = \begin{bmatrix} \mathbf{F}_{N_1,0} & \mathbf{F}_{N_1,1} & \cdots & \mathbf{F}_{N_1,na} \\ \mathbf{F}_{N_1+1,0} & \mathbf{F}_{N_1+1,1} & \cdots & \mathbf{F}_{N_1+1,na} \\ \vdots & \vdots & \vdots & \vdots \\ \mathbf{F}_{N_2,0} & \mathbf{F}_{N_2,1} & \cdots & \mathbf{F}_{N_2,na} \\ \mathbf{\Gamma}_{N_1,0} & \mathbf{\Gamma}_{N_1,1} & \cdots & \mathbf{\Gamma}_{N_1,nb-1} \\ \mathbf{\Gamma}_{N_1+1,0} & \mathbf{\Gamma}_{N_1+1,1} & \cdots & \mathbf{\Gamma}_{N_1+1,nb-1} \\ \vdots & \vdots & \vdots & \vdots \\ \mathbf{\Gamma}_{N_2,0} & \mathbf{\Gamma}_{N_2,1} & \cdots & \mathbf{\Gamma}_{N_2,nb-1} \\ \mathbf{\Theta}_{N_1,0} & \mathbf{\Theta}_{N_1,1} & \cdots & \mathbf{\Theta}_{N_1,nd-1} \\ \mathbf{\Theta}_{N_1+1,0} & \mathbf{\Theta}_{N_1+1,1} & \cdots & \mathbf{\Theta}_{N_1+1,nd-1} \\ \vdots & \vdots & \vdots & \vdots \\ \mathbf{\Theta}_{N_2,0} & \mathbf{\Theta}_{N_2,1} & \cdots & \mathbf{\Theta}_{N_2,nd-1} \end{bmatrix}$$

$$\mathbf{G} = \begin{bmatrix} \mathbf{G}_0 & 0 & \cdots & 0 \\ \mathbf{G}_1 & \mathbf{G}_0 & \cdots & 0 \\ \vdots & \vdots & \ddots & \vdots \\ \mathbf{G}_{N_p-1} & \mathbf{G}_{N_p-2} & \cdots & \mathbf{G}_0 \end{bmatrix}$$

$$\mathbf{G}' = \begin{bmatrix} \mathbf{G}'_0 & 0 & \cdots & 0 \\ \mathbf{G}'_1 & \mathbf{G}'_0 & \cdots & 0 \\ \vdots & \vdots & \ddots & \vdots \\ \mathbf{G}'_{N_p-1} & \mathbf{G}'_{N_p-2} & \cdots & \mathbf{G}'_0 \end{bmatrix}$$

$$\mathbf{G}_{N_{12u}} = \begin{bmatrix} \mathbf{G}_{N_1-1} & \mathbf{G}_{N_1-2} & \cdots & \mathbf{G}_{N_1-N_u} \\ \mathbf{G}_{N_1} & \mathbf{G}_{N_1-1} & \cdots & \mathbf{G}_{N_1-N_u+1} \\ \vdots & \vdots & & \vdots \\ \mathbf{G}_{N_2-1} & \mathbf{G}_{N_2-2} & \cdots & \mathbf{G}_{N_2-N_u} \end{bmatrix}$$

$$\mathbf{G}'_{N_{12u}} = \begin{bmatrix} \mathbf{G}'_{N_1-1} & \mathbf{G}'_{N_1-2} & \cdots & \mathbf{G}'_{N_1-N_u} \\ \mathbf{G}'_{N_1} & \mathbf{G}'_{N_1-1} & \cdots & \mathbf{G}'_{N_1-N_u+1} \\ \vdots & \vdots & & \vdots \\ \mathbf{G}'_{N_2-1} & \mathbf{G}'_{N_2-2} & \cdots & \mathbf{G}'_{N_2-N_u} \end{bmatrix}$$

$$\mathbf{G_p}(z^{-1}) = \begin{bmatrix} \mathbf{G}_{1p,0} + \mathbf{G}_{1p,1}z^{-1} + \cdots + \mathbf{G}_{1p,nb-1}z^{-(nb-1)} \\ \mathbf{G}_{2p,0} + \mathbf{G}_{2p,1}z^{-1} + \cdots + \mathbf{G}_{2p,nb-1}z^{-(nb-1)} \\ \vdots \\ \mathbf{G}_{N_pp,0} + \mathbf{G}_{N_pp,1}z^{-1} + \cdots + \mathbf{G}_{N_pp,nb-1}z^{-(nb-1)} \end{bmatrix}$$

$$\mathbf{H} = \begin{bmatrix} \mathbf{H}_0 & 0 & \cdots & 0 \\ \mathbf{H}_1 & \mathbf{H}_0 & \cdots & 0 \\ \vdots & \vdots & \ddots & \vdots \\ \mathbf{H}_{N_p-1} & \mathbf{H}_{N_p-2} & \cdots & \mathbf{H}_0 \end{bmatrix}$$

$$\mathbf{H}' = \begin{bmatrix} \mathbf{H}'_0 & 0 & \cdots & 0 \\ \mathbf{H}'_1 & \mathbf{H}'_0 & \cdots & 0 \\ \vdots & \vdots & \ddots & \vdots \\ \mathbf{H}'_{N_p-1} & \mathbf{H}'_{N_p-2} & \cdots & \mathbf{H}'_0 \end{bmatrix}$$

$$\mathbf{H}_{N_{12u}} = \begin{bmatrix} \mathbf{H}_{N_1-1} & \mathbf{H}_{N_1-2} & \cdots & \mathbf{H}_{N_1-N_u} \\ \mathbf{H}_{N_1} & \mathbf{H}_{N_1-2} & \cdots & \mathbf{H}_{N_1-N_u+1} \\ \vdots & \vdots & & \vdots \\ \mathbf{H}_{N_2-1} & \mathbf{H}_{N_2-2} & \cdots & \mathbf{H}_{N_2-N_u} \end{bmatrix}$$

$$\mathbf{H'}_{N_{12u}} = \begin{bmatrix} \mathbf{H'}_{N_1-1} & \mathbf{H'}_{N_1-2} & \cdots & \mathbf{H'}_{N_1-N_u} \\ \mathbf{H'}_{N_1} & \mathbf{H'}_{N_1-2} & \cdots & \mathbf{H'}_{N_1-N_u+1} \\ \vdots & \vdots & & \vdots \\ \mathbf{H'}_{N_2-1} & \mathbf{H'}_{N_2-2} & \cdots & \mathbf{H'}_{N_2-N_u} \end{bmatrix}$$

$$\mathbf{H_p}(z^{-1}) = \begin{bmatrix} \mathbf{H}_{1p}(z^{-1}) \\ \mathbf{H}_{2p}(z^{-1}) \\ \vdots \\ \mathbf{H}_{N_pp}(z^{-1}) \end{bmatrix}$$

$$= \begin{bmatrix} \mathbf{H}_{1p,0} + \mathbf{H}_{1p,1}z^{-1} + \cdots + \mathbf{H}_{1p,nd-1}z^{-(nd-1)} \\ \mathbf{H}_{2p,0} + \mathbf{H}_{2p,1}z^{-1} + \cdots + \mathbf{H}_{2p,nd-1}z^{-(nd-1)} \\ \vdots \\ \mathbf{H}_{N_pp,0} + \mathbf{H}_{N_pp,1}z^{-1} + \cdots + \mathbf{H}_{N_pp,nd-1}z^{-(nd-1)} \end{bmatrix}$$

$$\mathbf{HFG_pH_p} = \left[\begin{array}{cccc} \mathbf{H}_0 & 0 & \cdots & 0 \\ \mathbf{H}_1 & \mathbf{H}_0 & \cdots & 0 \\ \vdots & \vdots & \ddots & \vdots \\ \mathbf{H}_{N_p-1} & \mathbf{H}_{N_p-2} & \cdots & \mathbf{H}_0 \\ \mathbf{F}_{1,0} & \mathbf{F}_{1,1} & \cdots & \mathbf{F}_{1,na} \\ \mathbf{F}_{2,0} & \mathbf{F}_{2,1} & \cdots & \mathbf{F}_{2,na} \\ \vdots & \vdots & \vdots & \vdots \\ \mathbf{F}_{N_p,0} & \mathbf{F}_{N_p,1} & \cdots & \mathbf{F}_{N_p,na} \\ \mathbf{G}_{1p,0} & \mathbf{G}_{1p,1} & \cdots & \mathbf{G}_{1p,nb-1} \\ \mathbf{G}_{2p,0} & \mathbf{G}_{2p,1} & \cdots & \mathbf{G}_{2p,nb-1} \\ \vdots & \vdots & \vdots & \vdots \\ \mathbf{G}_{N_pp,0} & \mathbf{G}_{N_pp,1} & \cdots & \mathbf{G}_{N_pp,nb-1} \\ \mathbf{H}_{1p,0} & \mathbf{H}_{1p,1} & \cdots & \mathbf{H}_{1p,nd-1} \\ \mathbf{H}_{2p,0} & \mathbf{H}_{2p,1} & \cdots & \mathbf{H}_{2p,nd-1} \\ \vdots & \vdots & \vdots & \vdots \\ \mathbf{H}_{N_pp,0} & \mathbf{H}_{N_pp,1} & \cdots & \mathbf{H}_{N_pp,nd-1} \end{array} \right]$$

$$\mathbf{H'F\Gamma\Theta} = \begin{bmatrix} \mathbf{H'}_0 & 0 & \cdots & 0 \\ \mathbf{H'}_1 & \mathbf{H'}_0 & \cdots & 0 \\ \vdots & \vdots & \ddots & \vdots \\ \mathbf{H'}_{N_p-1} & \mathbf{H'}_{N_p-2} & \cdots & \mathbf{H'}_0 \\ \mathbf{F}_{1,0} & \mathbf{F}_{1,1} & \cdots & \mathbf{F}_{1,na} \\ \mathbf{F}_{2,0} & \mathbf{F}_{2,1} & \cdots & \mathbf{F}_{2,na} \\ \vdots & \vdots & \vdots & \vdots \\ \mathbf{F}_{N_p,0} & \mathbf{F}_{N_p,1} & \cdots & \mathbf{F}_{N_p,na} \\ \mathbf{\Gamma}_{1,0} & \mathbf{\Gamma}_{1,1} & \cdots & \mathbf{\Gamma}_{1,nb-1} \\ \mathbf{\Gamma}_{2,0} & \mathbf{\Gamma}_{2,1} & \cdots & \mathbf{\Gamma}_{2,nb-1} \\ \vdots & \vdots & \vdots & \vdots \\ \mathbf{\Gamma}_{N_p,0} & \mathbf{\Gamma}_{N_p,1} & \cdots & \mathbf{\Gamma}_{N_p,nb-1} \\ \mathbf{\Theta}_{1,0} & \mathbf{\Theta}_{1,1} & \cdots & \mathbf{\Theta}_{1,nd-1} \\ \mathbf{\Theta}_{2,0} & \mathbf{\Theta}_{2,1} & \cdots & \mathbf{\Theta}_{2,nd-1} \\ \vdots & \vdots & \vdots & \vdots \\ \mathbf{\Theta}_{N_p,0} & \mathbf{\Theta}_{N_p,1} & \cdots & \mathbf{\Theta}_{N_p,nd-1} \end{bmatrix}$$

$$\mathbf{T}(z^{-1}) = \mathbf{T}_0 + \mathbf{T}_1 z^{-1} + \mathbf{T}_2 z^{-2} + \cdots + \mathbf{T}_{nt} z^{-nt}$$

$$\tilde{\mathbf{U}} = \begin{bmatrix} \Delta\mathbf{u}(t) \\ \Delta\mathbf{u}(t+1) \\ \cdots \\ \Delta\mathbf{u}(t+N_p-1) \end{bmatrix}$$

$$\tilde{\mathbf{U}}_{N_u} = \begin{bmatrix} \Delta\mathbf{u}(t) \\ \Delta\mathbf{u}(t+1) \\ \cdots \\ \Delta\mathbf{u}(t+N_u-1) \end{bmatrix}$$

$$\tilde{\mathbf{V}} = \begin{bmatrix} \Delta\mathbf{v}(t+1) \\ \Delta\mathbf{v}(t+2) \\ \cdots \\ \Delta\mathbf{v}(t+N_p) \end{bmatrix}$$

$$\tilde{\mathbf{V}}\mathbf{YUV} = \begin{bmatrix} \Delta\mathbf{v}(t+1) \\ \Delta\mathbf{v}(t+2) \\ \vdots \\ \Delta\mathbf{v}(t+N_p) \\ \mathbf{y}(t) \\ \mathbf{y}(t-1) \\ \vdots \\ \mathbf{y}(t-na) \\ \Delta\mathbf{u}(t-1) \\ \Delta\mathbf{u}(t-2) \\ \vdots \\ \Delta\mathbf{u}(t-nb) \\ \Delta\mathbf{v}(t) \\ \Delta\mathbf{v}(t-1) \\ \vdots \\ \Delta\mathbf{v}(t-nd+1) \end{bmatrix}$$

$\tilde{\mathbf{V}}\mathbf{YUV}^f$ is defined accordingly via $\Delta\mathbf{v}(t+1)$, $\mathbf{y}^f(t)$, $\Delta\mathbf{u}^f(t-1)$ and $\Delta\mathbf{v}^f(t)$

$$\mathbf{W} = \begin{bmatrix} \mathbf{w}(t+1) \\ \mathbf{w}(t+2) \\ \vdots \\ \mathbf{w}(t+N_p) \end{bmatrix}$$

$$\mathbf{W}_{N_{12}} = \begin{bmatrix} \mathbf{w}(t+N_1) \\ \mathbf{w}(t+N_1+1) \\ \vdots \\ \mathbf{w}(t+N_2) \end{bmatrix}$$

$$\mathbf{Y} = \begin{bmatrix} \hat{\mathbf{y}}(t+1) \\ \hat{\mathbf{y}}(t+2) \\ \vdots \\ \hat{\mathbf{y}}(t+N_p) \end{bmatrix}$$

$$\mathbf{Y}_{N_{12}} = \begin{bmatrix} \hat{\mathbf{y}}(t+N_1) \\ \hat{\mathbf{y}}(t+N_1+1) \\ \vdots \\ \hat{\mathbf{y}}(t+N_2) \end{bmatrix}$$

$$\mathbf{YU} = \begin{bmatrix} \mathbf{y}(t) \\ \mathbf{y}(t-1) \\ \vdots \\ \mathbf{y}(t-na) \\ \Delta\mathbf{u}(t-1) \\ \Delta\mathbf{u}(t-2) \\ \vdots \\ \Delta\mathbf{u}(t-nb) \end{bmatrix}$$

$$\mathbf{YUV} = \begin{bmatrix} \mathbf{y}(t) \\ \mathbf{y}(t-1) \\ \vdots \\ \mathbf{y}(t-na) \\ \Delta\mathbf{u}(t-1) \\ \Delta\mathbf{u}(t-2) \\ \vdots \\ \Delta\mathbf{u}(t-nb) \\ \Delta\mathbf{v}(t) \\ \Delta\mathbf{v}(t-1) \\ \vdots \\ \Delta\mathbf{v}(t-nd+1) \end{bmatrix}$$

$\mathbf{YUV}^f$ is defined accordingly via $\mathbf{y}^f(t)$, $\Delta\mathbf{u}^f(t-1)$ and $\Delta\mathbf{v}^f(t)$

$$\mathbf{\Gamma}(z^{-1}) = \begin{bmatrix} \mathbf{\Gamma}_1(z^{-1}) \\ \mathbf{\Gamma}_2(z^{-1}) \\ \vdots \\ \mathbf{\Gamma}_{N_p}(z^{-1}) \end{bmatrix} = \begin{bmatrix} \mathbf{\Gamma}_{1,0} + \mathbf{\Gamma}_{1,1}z^{-1} + \cdots + \mathbf{\Gamma}_{1,nb-1}z^{-(nb-1)} \\ \mathbf{\Gamma}_{2,0} + \mathbf{\Gamma}_{2,1}z^{-1} + \cdots + \mathbf{\Gamma}_{2,nb-1}z^{-(nb-1)} \\ \vdots \\ \mathbf{\Gamma}_{N_p,0} + \mathbf{\Gamma}_{N_p,1}z^{-1} + \cdots + \mathbf{\Gamma}_{N_p,nb-1}z^{-(nb-1)} \end{bmatrix}$$

$$\mathbf{\Theta}(z^{-1}) = \begin{bmatrix} \mathbf{\Theta}_1(z^{-1}) \\ \mathbf{\Theta}_2(z^{-1}) \\ \vdots \\ \mathbf{\Theta}_{N_p}(z^{-1}) \end{bmatrix}$$

$$= \begin{bmatrix} \mathbf{\Theta}_{1,0} + \mathbf{\Theta}_{1,1}z^{-1} + \cdots + \mathbf{\Theta}_{1,nd-1}z^{-(nd-1)} \\ \mathbf{\Theta}_{2,0} + \mathbf{\Theta}_{2,1}z^{-1} + \cdots + \mathbf{\Theta}_{2,nd-1}z^{-(nd-1)} \\ \vdots \\ \mathbf{\Theta}_{N_p,0} + \mathbf{\Theta}_{N_p,1}z^{-1} + \cdots + \mathbf{\Theta}_{N_p,nd-1}z^{-(nd-1)} \end{bmatrix}$$

E.2.2 Dimensions

$$\mathbf{A}(z^{-1}) : \text{Matrix}(n \times n) \qquad \text{(rows} \times \text{columns)}$$
$$\mathbf{B}(z^{-1}) : \text{Matrix}(n \times m)$$
$$\mathbf{C}(z^{-1}) : \text{Matrix}(n \times n)$$
$$\mathbf{D}(z^{-1}) : \text{Matrix}(n \times l)$$
$$\mathbf{E}_j(z^{-1}) : \text{Matrix}(n \times n)$$
$$\mathbf{F}_j(z^{-1}) : \text{Matrix}(n \times n)$$
$$\mathbf{f}(t), \mathbf{f}'(t) : \text{Vector}(n \times 1)$$
$$\mathbf{G_d}(z^{-1}), \mathbf{G}_j(z^{-1}), \mathbf{G}_{jp}(z^{-1}) : \text{Matrix}(n \times m)$$
$$\mathbf{G}_j{}'(z^{-1}) : \text{Matrix}(n \times m) \qquad \text{See note!}$$
$$\mathbf{H_d}(z^{-1}), \mathbf{H}_j(z^{-1}), \mathbf{H}_{jp}(z^{-1}) : \text{Matrix}(n \times l)$$
$$\mathbf{H}_j{}'(z^{-1}) : \text{Matrix}(n \times l) \qquad \text{See note!}$$
$$\mathbf{T}(z^{-1}) : \text{Matrix}(n \times n)$$
$$\mathbf{u}(t), \mathbf{u}^f(t) : \text{Vector}(m \times 1)$$
$$\mathbf{v}(t), \mathbf{v}^f(t) : \text{Vector}(l \times 1)$$
$$\mathbf{w}(t) : \text{Vector}(n \times 1)$$
$$\mathbf{y}(t), \mathbf{y}^f(t) : \text{Vector}(n \times 1)$$
$$\mathbf{\Gamma}_j(z^{-1}) : \text{Matrix}(n \times m) \qquad \text{See note!}$$
$$\mathbf{\Theta}_j(z^{-1}) : \text{Matrix}(n \times l) \qquad \text{See note!}$$
$$\boldsymbol{\xi}(t) : \text{Vector}(n \times 1)$$

Note: If *filtering is applied* the vector of actuating variables $\mathbf{u(t)}$, the vector of disturbances $\mathbf{v(t)}$ and the vector of controlled variables $\mathbf{y(t)}$ must all have *the same dimension* (see chapter 8.3.3 on page 106 et seqq.). Therefore in all cases $n = m = l$ applies.

Appendix F

Methods for matrix inversion

For matrix inversion several algorithms exist whose suitability for a matrix inversion necessary for a GPC controller will be examined in the following. Special attention is given to the computation time that is necessary for the matrix inversion.

F.1 Gauss algorithm

The basic idea of the Gauss algorithm is to solve the matrix equation $\mathbf{A} \cdot \mathbf{A}' = \mathbf{I}$. The algorithm is described in the following:

1. First the matrix $\mathbf{A}$ has to be transformed into triangular form. For this purpose, each diagonal element is removed from the underneath matrix rows one after the other so that then, the matrix has a triangular form. The multipliers for each case are stored in an auxiliary matrix $\mathbf{B}$. Pivoting for the optimization of the calculation was not implemented.

2. After having transformed the input matrix $\mathbf{A}$, the same will also be carried out with the identity matrix $\mathbf{I}$ using the factors stored in matrix $\mathbf{B}$.

3. Since now $\mathbf{A}$ and $\mathbf{I}$ are both transformed in identical manner into $\tilde{\mathbf{A}}$ and $\tilde{\mathbf{I}}$, the equation $\tilde{\mathbf{A}} \cdot \mathbf{A}' = \tilde{\mathbf{I}}$ results, in which $\tilde{\mathbf{A}}$ is an upper triangular matrix. The identity matrix $\mathbf{I}$ has transformed into a lower triangular matrix $\tilde{\mathbf{I}}$ whose diagonal elements are all 1. The system of equations can easily be solved and then, the elements of the matrix $\mathbf{A}' = \mathbf{A}^{-1}$ are determined.

F.2 Gauss-Jordan algorithm

The algorithm of Gauss Jordan corresponds to the algorithm by Gauss, explained in section F.1, with the difference that the matrix $\mathbf{A}$ is not transformed into a triangular form, but into a diagonal form. This can be achieved if the diagonal elements are not only eliminated from the lower, but also from the above matrix rows.

F.3 Exchange algorithm

The exchange algorithm is based on the assumption that the equation $\mathbf{y} = \mathbf{A}\mathbf{x}$ can be written as $\mathbf{x} = \mathbf{A}'\mathbf{y}$ using the inverse matrix $\mathbf{A}' = \mathbf{A}^{-1}$. Therefore, the inversion of a matrix can be realized via the exchange of an independent variable with a dependent one. The procedure is very well explained by Ose et al. [98].

F.4 LR decomposition

The LR decomposition, also known as LU decomposition, decomposes the matrix $\mathbf{A}$ into a lower (left) and into an upper (right) triangular matrix $\mathbf{L}$ and $\mathbf{R}$, such that $\mathbf{A} = \mathbf{LR}$ applies. If pivoting is done before the decomposition, the exchange movements are stored in a permutation matrix $\mathbf{P}$ so that in the end $\mathbf{PA} = \mathbf{LR}$ applies. Therefore, $\mathbf{A} = \mathbf{P}^T\mathbf{LR}$ does apply, too. Since $\mathbf{L}$ and $\mathbf{R}$ are triangular matrices, they can be easily inverted and from $\mathbf{L}^{-1}$ and $\mathbf{R}^{-1}$ the inverse matrix $\mathbf{A}^{-1} = \mathbf{P}^T\mathbf{R}^{-1}\mathbf{L}^{-1}$ can be obtained. The exchange of $\mathbf{L}$ and $\mathbf{R}$ because of $(\mathbf{LR})^{-1} = \mathbf{R}^{-1}\mathbf{L}^{-1}$ has to be noted.

F.5 Algorithm of Cholesky

For reasons of completeness, the algorithm of Cholesky should be mentioned, too, which, compared to the LR decomposition, promises a significant reduction of the calculation complexity [112, 124]. But, since it is suitable only for the inversion of symmetrical matrices, it cannot be used for the calculation of MPC system matrices and so it is omitted from further discussions.

F.6 Computation times

For the implementation of GPC, it is extremely important to use very efficient methods for matrix operations. Therefore, for the programming of the methods described above, optimizations were carried out, i. e. all divisions with the same divisor within loops were replaced by allocations *before* and *multiplications within* the loop. Then the methods described above were tested. The results are presented in table F.1. For the examinations, a computer with an AMD Duron® processor (900 MHz) and 128 MB RAM was used.

	Matrix size				
Algorithm	3×3	10×10	15×15	25×25	100×100
Gauss	1.9	18.9	56.2	239.4	14390
Gauss-Jordan	1.1	11.8	35.5	395.1	26010
Exchange algorithm	1.2	19.9	58.7	268.6	16980
LR (with pivoting)	2.4	20.9	54.9	206.9	10010
LR (without pivoting)	1.8	16.1	45.6	177.0	9420

Table F.1: Calculation times for matrix inversion in µs

As the presented results clarify, the choice of the optimum method depends on the size of the matrix that should be inverted. For small matrices with sizes up to 15×15, the Gauss-Jordan algorithm is the fastest one, whereas for larger matrices, the LR decomposition is superior.

Appendix G

Alternative method for matrix decomposition

The method for decomposition of the transfer polynomial matrix $\mathbf{G_d}(z^{-1})$ into two left coprime matrices $\mathbf{A}(z^{-1})$ and $\mathbf{B}(z^{-1})$, originally developed by Goodwin/Sin [44, chapter 2.3.5] and explained by Camacho/Bordons [20, chapter 6.2.1], is based on the following approach:

At the beginning, $\mathbf{G_d}(z^{-1})$ is converted into an arbitrary right matrix fraction.

$$\mathbf{G_d}(z^{-1}) = \mathbf{N_R}(z^{-1})\mathbf{D_R}(z^{-1})^{-1}$$

Analog to the decomposition into a left fraction (see chapter 8.1.2 on page 85), $\mathbf{D_R}(z^{-1})$ is chosen as a diagonal matrix whose diagonal elements are equal to the smallest common denominator of the corresponding columns of $\mathbf{G_d}(z^{-1})$. Then, $\mathbf{N_R}(z^{-1})$ can easily be determined via $\mathbf{N_R}(z^{-1}) = \mathbf{G_d}(z^{-1})\mathbf{D_R}(z^{-1})$. Here, $\mathbf{N_R}(z^{-1})$ and $\mathbf{D_R}(z^{-1})$ do not have to be right coprime.

Now, a unimodular matrix $\mathbf{U}(z^{-1})$ has to be determined which fulfils the equation

$$\begin{bmatrix} \mathbf{U}_{11}(z^{-1}) & \mathbf{U}_{12}(z^{-1}) \\ \mathbf{U}_{21}(z^{-1}) & \mathbf{U}_{22}(z^{-1}) \end{bmatrix} \cdot \begin{bmatrix} \mathbf{D_R}(z^{-1}) \\ \mathbf{N_R}(z^{-1}) \end{bmatrix} = \begin{bmatrix} \mathbf{R}(z^{-1}) \\ 0 \end{bmatrix} \tag{G.1}$$

in which

$$\mathbf{U}(z^{-1}) = \begin{bmatrix} \mathbf{U}_{11}(z^{-1}) & \mathbf{U}_{12}(z^{-1}) \\ \mathbf{U}_{21}(z^{-1}) & \mathbf{U}_{22}(z^{-1}) \end{bmatrix}$$

$$\mathbf{P}(z^{-1}) = \begin{bmatrix} \mathbf{D_R}(z^{-1}) \\ \mathbf{N_R}(z^{-1}) \end{bmatrix}$$

$$\mathbf{R}(z^{-1}) = \text{Greatest common right divisor of } \mathbf{D_R}(z^{-1}) \text{ and } \mathbf{N_R}(z^{-1})$$

If $\mathbf{P}(z^{-1})$ is transformed into a upper right triangular form by elementary row transformations, the greatest common right divisor of $\mathbf{D_R}(z^{-1})$ and $\mathbf{N_R}(z^{-1})$ can be obtained according to the properties of polynomial matrices mentioned in appendix A. If the same transformations are applied to an identity matrix, the unimodular matrix $\mathbf{U}(z^{-1})$ will result.

It can be proven that the sub-matrices $\mathbf{U}_{21}(z^{-1})$ and $\mathbf{U}_{22}(z^{-1})$ are left coprime. Since $\mathbf{U}_{22}(z^{-1})$ is in addition to that non-singular, the following equation can be deduced from (G.1):

$$\mathbf{N_R}(z^{-1})\mathbf{D_R}(z^{-1})^{-1} = -\mathbf{U}_{22}(z^{-1})^{-1}\mathbf{U}_{21}(z^{-1})$$

A comparison with equation (8.5) results in

$$\mathbf{A}(z^{-1}) = \mathbf{U}_{22}(z^{-1})$$
$$\mathbf{B}(z^{-1}) = -\mathbf{U}_{21}(z^{-1})z$$

As already mentioned $\mathbf{U}_{21}(z^{-1})$ and $\mathbf{U}_{22}(z^{-1})$ are left coprime and thus a left coprime description is also found for $\mathbf{A}(z^{-1})$ and $\mathbf{B}(z^{-1})$.

The procedure described above is also described in detail by Geering [41] in German language.

Index